WOOD ENGRAVINGS OF BIRDS

Books by the same author

BRITISH NAMES OF BIRDS

BIRD ILLUSTRATORS

COLLECTING BIRD STAMPS

Beautiful or Paradise Parrakeet.

# WOOD ENGRAVINGS OF BIRDS

CHRISTINE
E. JACKSON

WITHERBY

First published in 1978
by H. F. & G. WITHERBY LTD.
5 Plantain Place, Crosby Row,
London SE1 1YN

ISBN 0 85493 128 7
Filmset in Monophoto 12pt Bembo
and printed in Great Britain by
BAS Printers Limited,
Over Wallop, Hampshire

WHITE STORK

The more I have since thought upon the subject, the more I am confirmed in the opinions I have entertained, that the use of woodcuts will know no end, or, so long as the importance of printing is duly appreciated and the liberty of the press held sacred.

Thomas Bewick.

QUEZAL

# *Contents*

WREN

# *Artists and Engravers Illustrated*

*Thomas Bewick*
*W. S. Coleman*
*Dalziel brothers*
*William Dickes*
*T. C. Eyton*
*Benjamin Fawcett*
*Alexander Fussell*
*Robert Gibbings*
*J. G. Keulemans*
*G. E. Lodge*
*A. F. Lydon*
*Mark of Wellington*
*Edward Neale*
*Charlton Nesbit*
*J. Thompson*
*Archibald Thorburn*
*C. F. Tunnicliffe*
*Charles Whymper*
*Josef Wolf*

HOATZIN

# *Acknowledgements*

I have received a great deal of assistance from many sources whilst preparing the text for my books, and am always pleased when it comes to the point where I can publicly acknowledge the help. I am particularly indebted to the staff of the Hertfordshire and Shropshire County Libraries, especially the Librarian in charge at Harpenden, Mr Eric Brandreth, and Mr M. F. Messenger at Shrewsbury. Mr R. Fish at the Zoological Society of London and Mr Rowlands at the British Museum (Natural History), and their staff, frequently go further than the normal call of duty and draw my attention to material I might otherwise have missed. To Miss J. M. Hermann a special note of thanks is due for the generous loan of so many bird books from the Special Collection at Newington Library in the London Borough of Southwark.

Mr R. C. G. Morris-Eyton of Newport, and Mr S. Morley Tonkin, gave me willing help in an attempt to trace the elusive engraver, Mr Mark of Wellington, from sources in Shropshire. That we all failed in this endeavour is certainly not for want of trying. My Canadian friend, Mr David Lank, has maintained a kindly interest in my efforts to trace the wood engravers of birds and his generous present of his own book on Bewick, *Once-upon-a-Tyne*, earlier this year, was a timely inspiration. Mr David McDonald Booth and Mr Peter Tate have also given me their help during the course of production of the book. Mr John Physick, of the Victoria and Albert Museum, kindly supplied me with details of the career of Richard A. Thompson at the South Kensington Museum.

To my publisher, Mr Antony Witherby, I am more than usually indebted. He very kindly undertook all the work on the illustrations whilst I was preoccupied with moving house. For this, and his many suggestions concerning the text, besides the meticulous care over the design of the book, I am especially grateful.

Bewick's toolbox is now in the care of the Newcastle-upon-Tyne Library and maintained just as Bewick left it. I am pleased that permission was granted by the Librarian to use the illustration of the tools Bewick had fitted up and adapted to his particular use for the engraving of his masterpiece *History of British Birds*, and acknowledgement is made to the City Engineer, Newcastle.

For many years I have admired the drawings, paintings and engravings of Mr C. F. Tunnicliffe and I was very delighted and gratified when he, and his publishers, Gollancz, gave their permission for three of his beautiful wood engravings to be reproduced in the final chapter. They form the perfect conclusion, since they confirm Bewick's belief that 'the use of woodcuts will know no end.'

INDIAN JAÇANA

BUSTARD

# *Introduction*

The art of wood engraving has a special place in the illustration of books on birds and natural history.

Many hundreds of attractive wood engravings of birds have been used from the time Bewick made wood engraved birds popular at the end of the 18th century up to the present day. A large proportion of these bird figures are good accurate portraits and many have background scenes of considerable charm and authenticity. Some of the engravers of the birds are well-known, whilst others remain in obscurity despite a good deal of effort to trace them. However, though the biographical details of the craftsmen may prove elusive, sufficient examples of their work remain to demonstrate both the remarkably high standard of their engraving and also to give us a great deal of pleasure today. Indeed, such is the popularity of wood engraving as a means of illustration, that of all the hand techniques, either using metal in engraving and etching, aquatint and stipple, or drawing on lithographic stones, wood engraving is the only one still used to any extent at the present time.

The use of wood for making prints has a very long history since it was the first material to be used for illustrations. The earliest known print, of any description, was made from wood blocks, taken by order of the Empress Shotoku of Japan (A.D. 762–769). Woodcuts were first used regularly in Europe in the early 15th century, and in England in the last quarter of the 15th century.

The woodcut differs in a number of points from a wood engraving. A woodcut is made on a block of soft wood with the grain running horizontally, by cutting away any area not required to print black. In other words, the wood cutter works the wood to produce a black line, or an area to print black. A wood engraving is produced by cutting a white line across the fine, smoothly polished end grain of hardwoods. When a print is taken, the difference lies in the finer, more detailed white lines obtained by using the close grain of hardwoods, as opposed to the thicker black areas of a woodcut.

Because the woodcut was incapable of reproducing fine details, it lost favour in competition with metal engraving. Wood was little used for over a century until Bewick demonstrated that illustrations from it could be as effective and detailed as from metal, provided a hard wood worked on the end grain was used. Bewick was a very good artist as well as a craftsman, and his designs were a revelation in their standard of artistic achievement as well as scientific accuracy. Unfortunately, Bewick and later engravers rarely signed their blocks, and authors of books in which their wood engravings appeared did not always state who their artists and engravers were. In consequence, we have great difficulty today knowing who painted the original water-colour design for the engraving; who copied or traced the design onto the wood block; and who incised the design into the block. Sometimes one person did all these processes, Bewick, Fawcett and J. Thompson were their own artists as well as engravers. Even then, their individual methods varied. For example, Bewick worked his own designs straight onto the block for many years until, in the early 1790s, he preferred to adopt the new style of painting in water-colours and made a preliminary water-colour sketch, or rather pen and ink drawing tinted with water-colour washes.

Fawcett and Thompson signed their blocks on a number of occasions 'Fawcett del et sc.' and 'Thompson del et sc.', and in these circumstances it is reasonably safe to assume that they were wholly responsible for the design and its engraving. Many of our great bird artists had their water-colour paintings and pencil or charcoal drawings copied on to the wood by wood engravers from the 1830s onwards. It was not uncommon for the artist to draw the bird onto a wooden block and then get it engraved by a craftsman in the printer's workshop. Where two sets of initials or signatures appeared on a print it usually meant that the artist's name was given in the bottom left hand section of the design, the name of the engraver appearing in the right hand corner. Occasionally the positions are reversed, or the initials centred.

The diversity of rôles is reflected in the chapter titles to this book. The most important and interesting books about birds which have wood engraved illustrations are described, and the people responsible for their being issued take their place at the head of the majority of chapters. Thus we have the artist-engravers Bewick, Dickes and Fawcett; and the two authors Eyton and Yarrell who employed engravers to execute the illustrations in their bird books.

Bewick, the source and inspiration of all the others,

takes pride of place as the first of these producers of bird books with wood engraved figures. When Bewick died in 1828 there were fewer than ten wood engravers at work on book illustrations, but by 1840 there were over fifty in London alone. William Yarrell, who emulated Bewick in almost every detail of format and even adopted Bewick's title, employed the best London engraver of his day in John Thompson. Alexander Fussell, Yarrell's artist, and Thompson were engaged on three editions of his book from about 1836 to 1856.

A wood engraver at work in the provinces in 1836 also executed some engravings very similar in style to those of Bewick. The work was done in Wellington, Shropshire, for T. C. Eyton who wrote a supplement to Bewick's book and drew the birds himself but employed 'Mr Mark' to engrave the blocks. The great Yorkshire artist-engraver and creator of coloured bird prints, Benjamin Fawcett, worked in the East Riding for sixty years up to the end of the 19th century, and at his death the era of printing in colour from wood came to an end. Because he made Bewick's acquaintance, whilst still a young man, he forms a link in the century long chain of eminent bird engravers from the 1790s to the 1890s.

Dickes is most probably the least well-known of these people, and besides choosing him on account of his many very good bird engravings, he also stands as representative for all the many long forgotten wood engravers who worked to a high standard of craftsmanship in order to produce relatively cheap books which instructed and entertained in a popular manner. All of these artists, authors and engravers, genuinely sought to inform the ever-increasingly literate population in the 19th century of a better way to appreciate nature, and to share with their readers the delight and pleasure which birds invariably give to successive generations.

# *History and Development*

The earliest prints of birds in British books most probably are the woodcuts in Bartholomew Anglicus' *De Proprietatibus Rerum* printed about 1495. They were very crudely cut and the birds consequently unidentifiable. A few more birds may be seen in Gervase Markham's two books *Hunger's Prevention* of 1621 and the 10th edition of his *Method or Epitome* describing remedies for diseases in birds and other domestic animals, published in 1671. Three woodcuts illustrated the section on birds and the title-page also had a number of small bird woodcuts. These were all primitive drawings and the woodcuts, due to the limitations of the process and materials then used, were rudimentary.

The only bird book with early woodcuts of any ornithological significance was *Ornithologia Nova*, 1743–5. The woodcuts are crude and of the 353 birds depicted not many of them are identifiable. One of the more attractive figures is that of the Ruff. The book includes British specimens as well as many foreign birds.

Woodcuts could not compete with the fineness of the intaglio techniques, copper-engraving and mezzotint, which were introduced in the 17th century to reproduce drawings and paintings. In consequence, woodcuts went out of favour in the 18th century, and throughout this period copper engravings and etchings were used for bird illustrations.

Within the space of two decades, copper's two great rivals, wood and stone, emerged on the printing scene. The first English book to have lithographic illustrations appeared in 1807 (J. T. Smith's *Antiquities of Westminster*) and seventeen years before this Thomas Bewick commenced the publication of his book about quadrupeds with its incomparable wood-engravings. For sixty years, all three materials were in use, but only wood has remained popular to the present day. Whilst copper plates ceased to be used in face of the competition from lithography, wood engraved plates flourished at the same time, both in monochrome and in colour. It is a curious fact, moreover, that the three great innovators of the use of wooden blocks (for engraved monochrome figures; for hand-coloured wood-engravings; and for colour printed plates) all made their names or firmly established their reputations with bird books. Thomas Bewick revived the art of engraving, but used the end grain of hard wood blocks, and his finest work is to be seen in *A History of British Birds* (1st edition 1797–1804). Benjamin Fawcett of Driffield used larger blocks for his plates of birds which were then hand coloured, the most notable title with this type of illustration being *A History of British Birds*, 1851–7 by the Reverend F. O. Morris. Fawcett also developed chromoxylography, or the printing in colour from a series of wood blocks, one engraved block for each colour. In 1834 the third craftsman, George Baxter, demonstrated his method of colour printing in Mudie's

*The Feathered Tribes of the British Islands.* The title page vignettes to the two volumes were engraved on wood and printed in colour by him, using water inks.

The great flourishing of wood-engraving was due to Thomas Bewick who introduced the process to Britain at the end of the 18th century and I have devoted a separate chapter to him. He not only revived the use of wood but added a new dimension when he used hard wood instead of soft wood. Bewick actually chose to engrave on the end grain of boxwood, though other hard woods such as maple, dried for a long time, may be employed. The surface was carefully ground smooth and then a 'white line' was cut with an engraving tool, the graver or burin.

The advantages of this technique lay in the freedom with which the artist could work on the wood and the ability of the hard wood to stand up to the pressure of the printing press for long runs. Bewick claimed that one of his blocks had been used for 900,000 impressions, but this was unusual, for the block often cracked in the printing process before that.

A major title to appear with monochrome wood-engravings was Yarrell's *A History of British Birds* 1837-41, modelled on Bewick's work. Many artists and engravers worked on the numerous wood-engravings in the four editions. Among the engravers were the Thompsons, Charles Whymper and the Dalziel Brothers. Details of the Thompson family and their work is included in the separate chapter on Yarrell.

The Dalziel Brothers were operating in London from 1839 to 1893. They were a Northumberland family and spent their boyhood in Newcastle where they were closely associated with Bewick's pupils and other northern artists and craftsmen. The father, Alexander, was an artist, and seven of his eight sons by Elizabeth Mills became artists by profession. Four of the sons constituted the firm Dalziel Brothers, which produced engravings and drawings, and published a good proportion of the wood-engravings in books between 1840 and 1880. Their most important work, from our point of view, was executed for Yarrell's *History of British birds*, but, generally speaking, their blocks for the *Illustrated London News* and *Punch* most probably brought them greater fame and financial reward.

Other authors who employed the brothers to engrave the figures for their books were John George Wood (1827–89) and John Christopher Atkinson (1814–1900). Wood was a cleric who also lectured on natural history and worked in an anatomical museum. He was more interested in writing natural history books of a popular nature than any of these other pursuits. He became famous in that capacity so that his numerous publications brought him and his publishers a good income. Whilst Chaplain of St Bartholomew's Hospital he wrote *The Illustrated natural history (*1853) which had one of its three volumes devoted to birds. The 494 fine wood-engravings throughout the book were the work of the Dalziel Brothers. The blocks were all shapes and sizes, some with a number of birds, others with a single bird figure. The engraving of the feathering, scenery and foliage was delicate and beautifully executed. The artists were W. S. Coleman, Edward Neale, H. W. Weir, Josef Wolf, Thomas W. Wood and Johann Baptist Zwecker, also William Harvey.

J. G. Wood's other book, *Homes without hands* had 110 Dalziel engravings and was published in 1864–5. Some of the illustrations included the birds as well as their nests. Of much greater charm was Atkinson's *Sketches in natural history*, 1860–1, with 82 figures of which 32 were of birds, with additional diagrams of feet, diving methods, and eggs. William Stephen Coleman (*c.* 1829–1904) had done

GREAT SPOTTED WOODPECKER

the lively drawings of the birds, with branches, leaves and sky as a setting for the figures.

Another wood-engraving firm which handled bird book illustrations was run by a father and two sons. Josiah Wood Whymper (1813–1903) taught himself to engrave on wood. He spelled his name Whimper on his early

GREAT WHITE COCKATOO

work, altering it to Whymper, the form which he considered to be the original spelling, in 1840. In this same year he took up water-colour painting and started to exhibit his work four years later. Also in 1840, his son Edward was born, then thirteen years later Charles was added to the family. Edward (1840–1911) trained as a xylographer. He and his father did the twenty wood-engravings for Wolf's *Life and habits of wild animals,* illustrated from designs by Josef Wolf. It was published in 1873–4. The two sons were great travellers as well as being interested in natural history. Edward published an account of his *Travels amongst the Great Andes of the Equator* and Charles wrote a book on *Egyptian birds.* They illustrated many other travel and natural history volumes. The most successful title for which Josiah Wood Whymper did the engravings, was C. A. John's *British birds in their haunts.* The 190 engravings were after drawings by Josef Wolf for the first edition of 1862. The engravings were similar in style to those of Bewick, and the birds were drawn to one-third of their natural size. Charles did a few of the engravings for Yarrell's later editions of

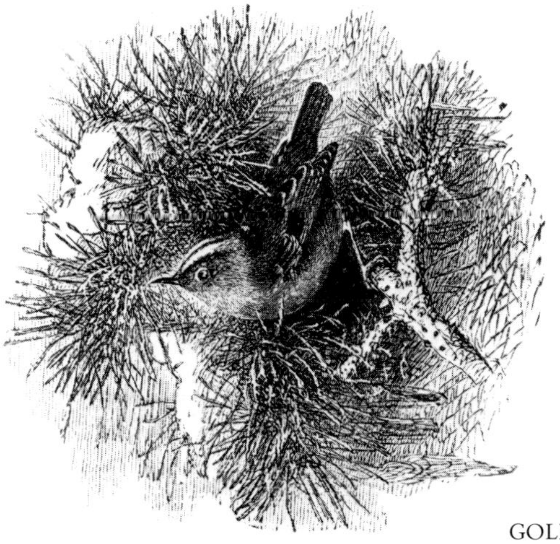

GOLD-CREST

*History of British birds.* Some of his most attractive engravings appeared in Henry Seebohm's two books *Siberia in Europe* in 1880 and *Siberia in Asia* 1882, which were reissued as one volume in 1901. These volumes have many attractive scenes, drawn on the expedition, and excellent wood engravings of the birds which they brought back with them. Ten of the 112 engravings included birds.

The value of finely detailed engravings for figures within the text was soon appreciated. Small engravings could be placed alongside the text at the precise place where they were needed to illustrate a point, since they could be printed at the same time as the type for the text. Where separate plates were used, they were often inserted some pages away from the point they illustrated. Wood engraved figures of small details, e.g. anatomical details, were used as supplements to the plates depicting a bird species. William Yarrell astutely adapted Bewick's use of the vignette for the purpose of reproducing anatomical (particularly osteological) drawings and these supplemented the main engravings of birds. William MacGillivray also used wood engraved anatomical details, but for a different reason.

MacGillivray had been bitterly disappointed at not being able to afford to illustrate his major work *A History of British Birds* with full-page metal engravings or etchings of his beautiful water-colour studies of birds. Instead, he illustrated his scholarly work with steel engraved plates of skeletons, and wood engraved figures of heads and feet in the main from his own drawings. The engraved figures are good of their kind, but the overall effect, with no illustrations of complete birds, was drab and unattractive. Nevertheless, the engravings showed how effectively wood could be used for reproducing careful and accurate drawings in some detail. The first volume of MacGillivray's five volumes appeared in 1837, the same year in which Yarrell's most attractive book of the same title also appeared. The beautiful engravings of birds in Yarrell's book ensured its success, whilst MacGillivray's work, spoiled by its lack of good illustrations, was largely ignored.

The lesson of the comparison between the illustrations in these two books was not lost on contemporary authors. Wood engraved vignettes of anatomical interest were a useful adjunct, but were not attractive as the sole means of illustration. Wood engraved anatomical figures were quickly adopted by authors whose books were illustrated in the main by full page lithographic plates. When lithography was first practised in the early 1830s, the illustrations produced were not finely drawn, so the sharper, clearer wood engraved figures were used for reproducing minutely lined drawings.

A number of the eminent lithographers had their drawings reproduced as text-figures to supplement their lithographs in bird books. William Swainson most probably did the designs for the 41 wood-engravings in *Fauna Boreali-Americana* (the second volume of which dealt with birds.) Swainson also included 300 engravings in his two volumes *On the natural history and classification of birds*, 1836–7. Josef Wolf had numerous designs copied by wood-engravers. Besides the twenty for Wood's *Illustrated natural history* he did 65 figures for Llewellyn Lloyd's *Game birds and wild fowl of Sweden and Norway*, a book which also described salt water fishes and seals. Wolf, C. Whymper, Keulemans and Thorburn all contributed drawings for James Edmund Harting's *Sketches of bird life*, 1883, which contained a number of delightful engravings as well as lithographs.

Archibald Thorburn, one of the best bird artists of more recent times, had a few other small drawings engraved for bird books, among them some very attractive little vignettes which were used to decorate W. Swaysland's *Familiar Wild Birds* 1883–8. These charming engravings were signed either with his initials or 'A. Thorburn', but the engraver's name did not appear. The engravings in Swaysland's book were ornamental, just as Bewick's tail-piece vignettes were intended to be decorative, and this was a further use to which small wood engravings were put in books containing plates by another medium. Decorative engravings were placed at the end of chapters, on title-pages and in what would otherwise have been empty spaces in the preliminary pages.

Henrik Grönvold was another artist who worked mainly for lithographers, but he also had drawings

PIED FLYCATCHER

GRASSHOPPER
WARBLER

reproduced by wood-engravers. He did an enormous number of drawings for W. L. Sclater and A. C. Stark's *The Birds of South Africa* in the early years of the 20th century. When the work was completed in 1906, the four volumes were illustrated by 500 figures.

The last group of wood-engraved text-figures of real merit, was done by the great wildfowl and mammal artist John Guille Millais (1865–1931). His work appeared in many books after 1900, most of his drawings being reproduced photographically. Twenty-eight of the wood-engravings of feathers, wings, heads, etc., for Henry Seebohm's *Geographical distribution of the family Charadriidae*, 1887–8, were engraved from his drawings. His own book of *Game birds and shooting sketches* had a number of small wood-engravings in the text signed 'J. G. M.' when published in 1892 and a second edition of 1894.

George E. Lodge was a friend of Millais and collaborated with him in the drawing and engraving of the illustrations for *Shooting: moor and marsh* (by Lord Walsingham and Sir Ralph Payne-Gallwey) in 1909. A number of the engravings in this book are signed by Lodge either as the artist and engraver, or as the engraver of Millais' and Charles Whymper's bird designs. Lodge worked on hundreds of small drawings and engraved quite a number of them himself for books published at the end of the 19th century. He had been apprenticed to a wood engraver following his training as an artist, and a large number of his very good engravings may be seen in Howard Saunders' *An Illustrated Manual of British Birds*, published in 1889; W. H. Hudson's *British Birds* of 1895; A. H. Evans' *Birds*, 1899, which formed volume IX of the Cambridge Natural History series, (a selection from this work is included in the first eight pages of this book); and some engravings which he had also designed in W. T. Greene's *Birds of the British Empire* of 1898.

POMATORHINE SKUA

# *Thomas Bewick 1753–1828*

When reviewing Bewick's life and achievements, it is necessary to consider his drawing, engraving, technical innovations, writing, and natural history interests. We have an artist-naturalist and artist-craftsman, in the same person.

Bewick was born in August 1753, in a cottage at Cherryburn, which he described in his *Memoir* as being 'situated on the south side of the Tyne, in the county of Northumberland, a short distance from the river.' His father was a farmer and he also owned a small colliery. The varied countryside around the farm had woods and fields, streams and meadows, moors and commons, all of which harboured many flowers, trees, insects, animals and birds. Bewick soon became fascinated by all of these, particularly the birds and animals, and he also showed a keen interest in fishing and shooting.

He was not fond of school, and frequently played truant. His habit of filling every spare inch of space on his slate and books with drawings of animals, scenes and birds, did not endear him to his masters. When taking a day off from school, he took advantage of the flat surfaces on the gravestones and the nice smooth flags of the church porch and the hearth at home, for drawing purposes. He used bits of chalk for these drawings, and was delighted when a friend thoughtfully gave him some paper and he discovered that using a pen and ink and blackberry juice 'made a grand change'. 'These,' he said. 'were succeeded by a camel-hair pencil and shells of colour', and with them he continued to draw direct from nature. He never had a drawing lesson in his life, unless one counts the instruction he received during the early days of his apprenticeship on how to copy the designs in Copeland's *Ornaments*. His passion for drawing influenced the choice of his future career. He wished, above all else, to be a pupil in some trade where he could spend most of his time drawing. By chance, a Newcastle engraver visited Thomas' godmother at Bywell, and was told of the boy's talents and this was how he came to be apprenticed.

At the age of fourteen he was apprenticed to Ralph Beilby, a craftsman through whose workshop passed very varied work. Beilby came from a gifted family. His brother William and sister Mary were expert at painting glass with enamel colours and virtually introduced this craft to England. Ralph had worked as a jeweller with his father and was greatly esteemed as a very fine silver engraver. His workshop was in St Nicholas' churchyard, in Newcastle, and there Bewick joined him on the 1st October 1767, to start seven years' training in a diversity of crafts. The workshop produced seals, coffin- and door-plates, invitation cards, clock-faces and metal moulds into which glass bottles were blown. In addition, Newcastle had at this time a flourishing printing industry which produced a number of children's books containing

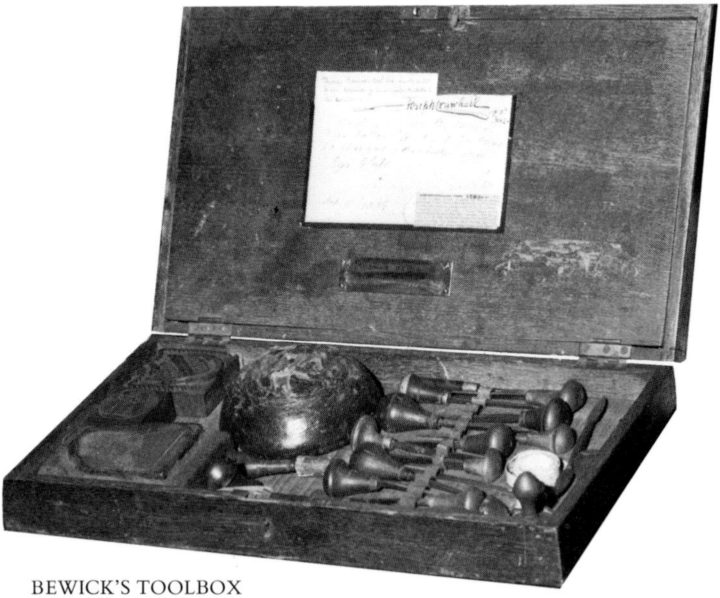

BEWICK'S TOOLBOX

woodcuts, many of which were ordered from Beilby. Some of Bewick's earliest examples of this type of work appeared in an illustrated alphabet *New lottery book of birds and beasts for children*, in 1771. Apart from learning all the other sides of the business, Bewick found these particular orders coming his way more frequently, since Beilby disliked working with wood and was not very proficient at wood-cutting.

Beilby 'fitted up and tempered his own tools, and adapted them to every purpose', teaching Bewick to do the same. This critical attitude to his tools was to enable Bewick to conceive a number of innovations and constantly seek to improve both the tools and the method of using them. In Bewick's estimation Beilby was the best kind of master an apprentice could have because he expected a very thorough knowledge of every part of the business to be gained by his pupils.

Towards the end of his apprenticeship, Bewick did a cut for a bar bill of George and the Dragon, and this was so much admired that more of this kind of work was brought to the workshop. Quite naturally, Bewick had shown over the seven years that his real talent lay in the design and execution of wood blocks. When the first of October 1774 arrived, Bewick was free, but he stayed at Newcastle a little while longer. During these few extra weeks he was paid a guinea a week.

Beilby had submitted some of Bewick's woodcuts for Gay's *Fables* in a competition for which he won a prize of seven guineas from the Society for the Encouragement of the Arts. The prize was for a cut called the Old Hound. Of this prize money Bewick said, 'I never in my life felt greater pleasure than in presenting it to my mother.'

Whilst an apprentice Bewick had formed the habit of walking the twelve miles back to Cherryburn to visit his parents every week. This he continued, as long as he worked in Newcastle and his parents were alive. He exulted in his physical toughness and excellent health. Sometimes he did the return journey on the same day, but on other occasions he stayed overnight and took time off to enjoy a few hours' fishing. On the walks he stored his memory with scenes of Tyneside and also took to thinking over political, philosophical and religious

concepts. He came to decisions concerning his attitudes to life and his behaviour towards other people, and firmly adhered to them during the rest of his long life.

In November 1774 he left Beilby and returned to Cherryburn to spend the winter at home. He had sufficient work to do, supplied by Newcastle printers. He enjoyed the freedom of the next few months and renewed his acquaintance with friends and neighbours, also with the other inhabitants of his valley. The 'robins, wrens, blackbirds, sparrows, a solitary crow' (i.e. hooded crow) were his 'intimate acquaintances', whilst he felt 'an extreme pleasure and curiosity in seeing the more rare visitants—such as the woodcock, the snipe, and waders, with the redwings, fieldfares, etc' when they made their appearance.

SPARROW

Almost on the spur of the moment, Bewick set out on a walking tour of Scotland in June 1776. He got as far as Stirling and visited Edinburgh, Glasgow and Dumbarton. He liked the people and found them friendly, including the craftsmen whom he visited and watched at work. From Edinburgh he took a sea-passage back to Newcastle. There he did sufficient work to pay his way to London where he proposed setting up in business.

He reached London on 1st October 1776 and proceeded to make contacts with friends who had moved south before him. Among these was a book-binder, William Gray, and a printer Robert Pollard, who saw to it that plenty of work reached his bench. Pollard had got permission from his master, Isaac Taylor, to pass on the work to Bewick, and Taylor was impressed by the young craftsman. Bewick, however, took a great dislike to both London and Londoners. He called them all 'Cockneys' and they in turn dubbed him a 'Scotsman'. He loathed the all too ample evidence of poverty and distress and felt he could never live amongst it and retain his earnings since he felt compelled to relieve any suffering which he saw. In addition, he could see little future for himself in London as a wood-engraver, because only copper engravings were in fashionable demand.

Taylor was most disapproving when he heard of Bewick's decision to return north, and broke off their relationship, but another printer, Thomas Hodgson, was more sympathetic and promised to send him work, and began by giving him sufficient to keep him busy for two years. This was a most acceptable arrangement to

DUNLIN

Bewick, because he did not wish to return to Newcastle to work in rivalry with Beilby for whom he had the greatest respect. With immense relief, Bewick returned home in June 1777, and started in Newcastle on his own.

No sooner was Bewick settled once more at his old lodgings 'at Hatfields' with his workbench set up there, than Beilby offered him a partnership in his firm. Bewick hesitated for some time because he liked both the freedom of choosing what work he would do and in working for himself. However, he decided to accept Beilby's offer, and went into the business, still in St Nicholas' churchyard, taking his brother John in with him as his apprentice. He was never sure that he had made the right choice. Certainly it meant that he left behind him fewer exquisite wood-engravings than he might otherwise have done. The reason for this is that almost all the engraving for his three outstanding books (*Quadrupeds, Birds, Fables*) was done in his spare time during the evenings when his day's work at the workshop was completed. That work was frequently other crafts than engraving on wood.

Between 1779 and 1790, when *Quadrupeds* appeared, Bewick was endeavouring to improve his technique and overcome the difficulties in getting the press-men to

LONG-EARED OWL

make good impressions from his blocks. The press-men tended to over-ink the blocks in such a way as to blurr and overlap the edges. Bewick countered this to some degree by shaving down the edges of the block. He next lowered the parts he wished to appear lighter in the finished print. One day he was shown a wood-cut by Albrecht Dürer and noted that cross-hatching on this, and on other woodcuts he had seen, was used to simulate shade. Bewick tried this, 'endeavouring to provide the like effect on my blocks, on the end way of the wood.' Having discovered how it was done, Bewick then abandoned the idea, since 'I never could discover any additional beauty or colour that the crossed strokes gave to the impression, beyond the effect produced by plain parallel lines.' He went on to explain that 'the thinnest strokes cut upon the plain surface will throw *some light* on the subject or design; and, if these strokes are made wider and deeper, it will receive more light' and so on, until much thicker lines, of equal thickness with the black line, will produce a grey shade. Yet further thickening of the strokes produces white. Bewick also revealed, 'The first time I have ever heard anything about colour being produced by plain engraving was in the compliments paid me by Dr Thomas Stout, for my engraving on his large silver box . . . and from that time I attempted *colour* upon the wood.' Here Bewick used the word 'colour' where we should use 'tone' and he also clearly spells out the essential difference of his method of engraving, from wood-cutting, for the engraver cuts the wood to produce white lines, whilst the wood-cutter worked to produce black lines. Only very rarely did Bewick use crossed lines or cross-hatching in his subsequent work.

Another result from lowering the surface of the block, was to produce an effect of distance. In addition, the cuts made in the lowered portion were better able to stand the pressure in the press than lighter, delicate lines on the surface of the unshaved block used to give the same effect. Bewick's blocks stood up to a great deal of use. Apart from the way in which the engraving had been done, the type of wood which Bewick used was a factor in his

getting many impressions from the same block. Before Bewick, most craftsmen working with wood had used softwood blocks such as pear, cutting the block along the plank, i.e. with the grain running horizontally in the block. Bewick worked with hardwoods, box in particular, and used a block with the grain running vertically. The grain of boxwood is both close and even, allowing fine engraving. Bewick's recognition of the qualities of the grain of box enabled him to do work which rivalled the fine engraving on copper which the metal-engravers were then producing. He went a stage further. Instead of copying designs from other artists, he drew his own original designs on to the wood. In doing this, he thought of his design in terms of wood-engraving, with figures and scenes interpreted by graver strokes on the wood, so that they were perfectly adapted to the medium by which they were reproduced. Had Bewick not been such a good artist, he would not have been so successful. Sometimes he worked straight onto the wood, working from memory and producing minute details from scenes of rural life, and engraving everyday occurrences in the village and farmyard. At other times, he made watercolour drawings of the subjects and then, from these drawings, designed his wood-engravings.

'Having from the time that I was a schoolboy been displeased with most of the figures in children's books, and particularly with those of the *Three Hundred Animals* . . . I at last came to the determination of making the attempt.' The result, Bewick's *History of Quadrupeds,* first demonstrated on a grand scale his mastery of the techniques which he himself had devised. Bewick said of it, 'Such animals as I knew, I drew from memory on the wood; others which I did not know were copies from Dr Smellie's *Abridgement of Buffon,* and other naturalists, and also from the animals which were from time to time exhibited in shows. Of these last, I made sketches first from memory, and then corrected and finished the drawings upon the wood from a second examination of the different animals.' These were travelling menageries which visited the villages and attended the many fairs held in the country. Bewick not only took advantage of their appearance at Newcastle, but also did some work occasionally for the proprietor of the menagerie. In 1799 he did four large engravings for one of these gentlemen, of an elephant, tiger, lion and zebra. Bewick started his work on the *Quadrupeds* on the 15th November 1785, the same day on which his father died. He found some consolation in work, and he needed some diversion, for this was the third member of his family to die that year, the other two being his mother and eldest sister.

*History of Quadrupeds,* with a text by Beilby, was finished and published in 1790. Bewick and Beilby did not issue their books in parts, as was usual at that time, but preferred to wait until all the engravings had been done and the text compiled, and presented them together, selling their book when completed. Most probably this was because they fitted the work on *Quadrupeds* around their other work, in the evenings and at odd moments so as not to interfere with the flow of work of the business.

The success of *Quadrupeds* was immediate and beyond their expectations. They sold 1600 copies within a year, and further issues appeared at later dates. This is quite understandable, for they were far better than any other figures engraved on wood and published before that date. Where Bewick had seen the living animals, and those domestic animals with which he was familiar, his engravings are excellent. The figures for the animals of which he knew little and had drawn from stuffed figures, were noticeably less realistic and accurate. The tail-pieces were charming, and beautifully executed, being the first significant use of the vignette in Bewick's work.

Some of the stuffed specimens which Bewick had used were in the Wycliffe Museum of Marmaduke Tunstall. Tunstall had succeeded to his family's estates in 1760 and resided for a number of years in London after finishing his education. He was a keen naturalist, a correspondent of Linnaeus, and the author of *Ornithologia Britannica* in 1771, (an early checklist). In this same year he was elected a Fellow of the Royal Society. Five years later he married and moved to his house at Wycliffe, on the Tees near Barnard Castle. A little while later, he moved his museum collection up to his Yorkshire home and lived a quiet country life. He came into contact with Bewick through a mutual friend, George Allan a celebrated antiquary and topographer. Tunstall and Bewick became correspondents, but seem never to have met. Tunstall offered to let Bewick draw from any specimens in his museum.

PEACOCK

George Allan had some further influence on Bewick. He lent him a number of natural history books, among them Albin's *A Natural History of Birds*, Belon's *L'Histoire de la nature des oyseaux*, Ray & Willughby's *Ornithology*, and Gesner's *Natural history*. It was through Allan, also, that Bewick came to do the magnificent engraving of the Chillingham bull for Marmaduke Tunstall.

Bewick's success with the *Quadrupeds,* his increasing interest in natural history, and his love of birds, plus his excitement over Edwards' *Natural History of Uncommon Birds* and White's *Selborne*, were all leading him to the point where he, too, wanted to publish a book about birds. He declared that Pennant, more than all the other writers 'opened out the largest field of information and on his works I bestowed the most attention.' Pennant's *British Zoology* had etched figures, and no doubt Bewick felt there was much room for improvement on those figures done by Mazell who was not an ornithologist. And so, in 1791, he began work on the figures for *A History of British Birds.*

'At the beginning of this undertaking I made up my mind to copy nothing from the works of others, but to stick to nature as closely as I could.' Accordingly Bewick went back to Tunstall's Museum to draw some of the bird specimens. Tunstall had died in October 1790, leaving his estates to his half brother William Constable of Burton, who died only six months later. Constable left the property to his nephews Edward and Francis Sheldon in succession. 'Having been invited by Mr Constable, the then owner of Wycliffe,' Bewick set off 'from Newcastle on 16th July 1791, and stayed for nearly two months', drawing from the stuffed specimens. He lodged at the house of John Goundry, who had preserved the birds for Tunstall. About 35 of Bewick's wood-engravings for the 1797-1804 edition of *A History of British Birds* were based on specimens in this collection. When Bewick wished to refer to the collection again, he was able to do so at the Grange, Darlington, the home of his friend George Allan, from Christmas time 1791 onwards. Allan bought the natural history museum, stated to have cost Tunstall £5,000 for the birds alone, from Edward Sheldon. Later still, this famous collection was to form part of the Newcastle Museum which was greatly enriched by another Bewick acquaintance John Hancock.

WHITE-TAILED EAGLE

Some of Bewick's drawings from these museum specimens were in water-colours. His first water-colour study for *British Birds* was the sea eagle which had 'Wycliffe 1791' engraved at the base of the block. As soon as he got back to Newcastle, Bewick began to engrave from the drawings of the species he had studied at Wycliffe. Further assistance was given to him by his sporting friends and other naturalists who sent him stuffed or dried birds. But, unlike so many artists, he soon recognised the limitations of this method of working and commented 'But I had not been long thus engaged till I found the very great difference between preserved specimens and those from nature; no regard having been paid, at that time, to fix the former in their proper attitudes, nor to place the different series of the feathers so as to fall properly upon each other. It has always given me a great deal of trouble to get at the markings of the dishevelled plumage; and, when done with every pains, I never felt satisfied with them. I was on this account driven to wait for birds newly shot, or brought to me alive, and in the intervals employed my time in designing and engraving tail-pieces or vignettes.'

CRESTED TITMOUSE

Gradually, the number of engravings, done at the end of the day's work, mounted up. Beilby wrote the text, mainly copying the information from Pennant and

WILLOW WREN

others, until all the figures and descriptions of land birds for the first volume of *British Birds* were ready for publication. Then Bewick received a nasty shock. He discovered that Beilby proposed to put his own name on the title-page as the author of the book, and to refer to Bewick, in terms of unqualified praise, only in the introduction. They got into such disagreement over this arrangement that they had to go to the length of setting up an arbitrating committee, drawn from their mutual friends, and agreed to accept the committee's decision. The committee adjudicated that the matter should be left without either of them as the author, the title-page merely stating 'Printed for Beilby and Bewick by Edward Walker of Newcastle'. Bewick was thoroughly upset over this and before he got the second volume, on water birds, ready for the press, he bought up Beilby's share in *Quadrupeds* and his share in *British Birds* for a total sum of £400. Bewick then had to supply his own text for volume II, which delayed its appearance. The text, again plagiarised, is of no significance today, and at the time of publication was not of any great merit. Beilby left the business soon after this, and Bewick had much more work on his hands, some of which he had first to re-learn, such as the fine engraving of bank notes and cheques.

*Water Birds*, the second volume of *History of British Birds*, was published in 1804 with 240 wood-engravings. The first volume had included 208 engravings and had been issued in three different formats: 1,000 copies demi-octavo at half a guinea; 850 thick and thin royal at 15/- and 13/- respectively; and 24 copies in imperial at a guinea. When the second volume appeared, the prices were higher—12/-; 15/-; 18/- and £1-4-0d. In view of Bewick's strictures on unsatisfactory press-work, it is interesting to note that in 1800 he printed 24 copies of *Land Birds* on imperial paper with his own hands in an attempt to avoid the excessive inking of his blocks which distorted and spoiled the impression taken by his printer.

These two volumes were Bewick's greatest achievement. They combined excellent craftsmanship with the perfect marrying of type and engravings on the page. They also showed a deep knowledge of, and sympathy for, the subjects of his figures. Bewick's reproduction of the texture of the feathering was unique, also his detail of the plumage and the arrangement of the feathers. His figures are, in the main, accurate. Fault may be found with those figures of birds with which he was not familiar, but his robin, magpie, thrushes, ducks and tits are lively, faithful representatives of their kind. The spotted winter coat of the starling seems to shimmer like the irridescent feathers of its live counterpart. All the feathers of the male blackbird are conveyed by skilful variation in the type of graver mark chosen for each part of the plumage. The wood-engravings were the first really accurate and true pictures of our most common birds, and set a new standard in their illustration.

The bird figures were imposed on small background landscape scenes, representing the birds' natural habitats. The corncrake is skulking in a stony field with flowers, grasses, and other plants in the background. Bewick had

CORN-CRAKE

kept this bird in his garden for a time to be able to draw it after studying it, and its habits, closely. The mute swan, with wings partly raised, floats serenely on the water, with a river bank closely planted with trees and shrubs behind it. Occasionally Bewick used a simpler setting, such as the branch of a tree with only a couple of leaves, on which is perched a great tit, Bewick's 'The Titmouse, Oxeye'.

The figures with elaborate backgrounds, like a robin in the wintry scene, stand out clearly with a three-dimensional effect. The snow in the robin's landscape does this naturally for this bird, but Bewick achieved the same effect when he had dense, dark backgrounds, either of buildings or foliage. In order to make the birds stand out in relief against this kind of background, Bewick engraved a fine white line to separate the outline of the bird figure from its surroundings.

The figure of the species was printed at the head of a new page, followed by the text describing that species and giving other information as to where it had been found, if a rarer species, with quotations from earlier authors. At the conclusion of the text, a small, usually horizontal oval, tail-piece was printed. The vignettes did not always include birds, but were, in the main, scenes of rural life. Many of them show the villagers shooting and

fishing, boys playing and men working, with the odd barnyard scene. They are charming, and show Bewick's intensely keen observation not only of the animals and birds of the countryside, but also of his fellow human beings. His view was not picturesque. Quite a few of the men in the vignettes have wooden legs. His countrymen are poverty-stricken, blind, drunk, and lame. More often than not, the wind was blowing, and hats flew from heads, whilst the villagers bent their bodies in the struggle against the strong wind. Lots of dogs appear, as companions, assistant wildfowlers, the subject of boy's pranks, or blind men's friends.

Of the few vignettes concerned with birds, one is a fine study of a single feather, and another shows a very broody mallard sitting under an overhanging tussock of grass, whilst a third shows a tawny owl which had found a dark place in the shadow of a tree canopy. The last two have a great deal of atmosphere. Within these small, restricted areas, sometimes as little as 40mm × 50mm, every part of the wood within the area of the picture is used and forms an integral part of the complete design.

Some of the work on the tail-pieces was done by Bewick's apprentices, but as he did not give them credit for their work and in no other way noted which parts he had done himself (Bewick hardly ever signed his own bird figure blocks), the extent of their contribution is unknown. Bewick had a number of pupils throughout his time in business, over forty years. Some of the engraving for the vignettes in *British Birds* is said to have

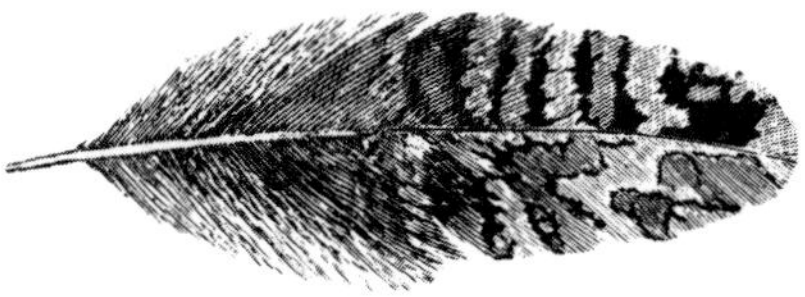

been executed by Robert Johnson, Charlton Nesbit, Luke Clennell and Henry Hole. Bewick's brother, John, was no longer with him at the time. He was Bewick's pupil for about five years and then went to London, and died whilst still a bachelor, in 1795. Bewick took his son Robert Elliot Bewick into partnership in 1812.

BITTERN

In his *Memoir*, Bewick mentions a few of the pupils and particularly commended Willian Harvey as being 'one of the first in excellence . . . who both as engraver and designer, stands pre-eminent at this day.' Harvey was apprenticed to Bewick from 1810 to 1817 and Bewick's esteem was shown in a kindly gesture on the first of January 1815 when he sent Harvey a copy of *British Birds* as a New Year's gift. Harvey's only contribution to a bird book, however, was as one of the artists who produced drawings of birds, to be engraved by others, for H. G. Adams' three volume *Our Feathered Families* published in 1862–3. His fellow artists included Luke Clennell (who was apprenticed to Bewick 1797–1804, but later abandoned wood-engraving for painting) and William Stephen Coleman. Between them they produced over 150 designs. Drawings by these three artists were used again for J. G. Wood's *Illustrated Natural History* 1851–3 (vol. 2. Birds). More of Harvey's delightful illustrations may be seen in E. T. Bennett's *The Tower Menagerie* of 1829. It is sad that his pupils did not follow Bewick's example and work more on natural history books. Bewick's influence was felt in many other fields, however, and was exerted by those of his pupils whose work was centred in London. He had revealed the previously unrealised potential of wood for illustrating books cheaply, even in competition with copper-engravings, at a time when the demand for cheap illustrated books was growing. This demand was to increase at an accelerated pace throughout the 19th century.

*British Birds* was so popular that Bewick saw six editions through the press in his own lifetime, and issued other smaller versions. He continued to add new figures and improve on the old ones, but many of the blocks were used time and again throughout successive editions. The sixth edition of 1825 is regarded as the best and is used as the standard edition for reference. It was Bewick's final attempt to make the engravings as nearly perfect as possible. During the course of issuing the later editions, Bewick also kept adding notes to the text. He up-dated information regarding the appearance in England of rare species and also often gave a glimpse of 19th century attitudes to, and use of, birds. He gives many details of thousands of geese driven from Suffolk to London to the markets, of the plucking of the geese for their feathers and quills, and of the ducks taken by decoy in the fens of Lincolnshire. He frequently deplores this exploitation of birds. In quoting Pennant's earlier account of the lark, 'the neighbourhood of Dunstable is famous from the great numbers of these birds found there, and that 4000 dozen have been taken between September and February for the London market', Bewick adds with some asperity, 'We must here dismiss the disgusting task of noting the edible qualities of these tiny creatures, the ornaments of our fields, our gardens, and groves; nor can we help regarding their destruction for the purposes of gormandism, as not a little reproachful to humanity in countries abounding with every species of food fit for the use of man.'

One interesting footnote tells of a very early ringing attempt. Mr E. M. Pleydell of Whatcombe, Dorset, was handed a live woodcock by his gamekeeper. It had become entangled in a net set to catch rabbits, during the winter of 1797. Mr Pleydell 'scratched the date upon a bit of thin brass, and bent it round the Woodcock's leg, and let it fly. In December the next year, Mr Pleydell shot this bird with the brass about its leg, in the very same wood where it had been first caught by the gamekeeper . . .'

In a letter to his friend John F. M. Dovaston of

Westfelton, Shropshire, when Bewick was at work on the sixth edition, he wrote, 'I am happy however to tell you that notwithstanding all difficulties I have nearly *fagged* through my task amounting to thirty-three new birds and shall gladly apply myself again to my *Tale* pieces, as I find it much less laborious to throw off those fancies than to *try* to put life into dead skins and ill-stuffed *Stittata* larks etc—what with strikes among the printers and the cold weather, the progress of the new Edition has been vexatiously retarded—and will not be out till after Christmas.' It was Dovaston who had made the pun on the tail-pieces and called them 'tale-pieces' since, he said, they usually told a story. Dovaston had gone on a pilgrimage to Newcastle in order to make Bewick's acquaintance and had found Bewick in the Blue Bell, one of his favourite taverns. This was 1st October 1823 and their lively correspondence began soon after and continued to within days of Bewick's death.

For the *Land Birds* 1797 Bewick had relied on the Wycliffe Museum, his own observation and his close friends for his subjects. Once that volume had been published, his number of correspondents grew and many more people, now aware that he needed specimens or anxious to contribute a 'new' bird, sent him stuffed birds for inclusion in *Water Birds* and the later editions of his book. By the time the sixth edition was published in 1825, the number of contributors had grown considerably. Bewick did not publish a 'list of contributors' but unfailingly recorded their help in the text.

LONG-
LEGGED
PLOVER

The water birds had been particularly difficult to obtain. Being so near the north-east coast, one would think that it would have been an easy matter to get hold of such common species as guillemots and puffins, and yet Bewick's records show that he was grateful when someone presented him with a specimen. His difficulty was almost certainly due to his own determination not to kill any living creature, thus preventing him from going out and shooting his own birds for specimens. Not once do we find a note that he had shot a bird. Nor in his letters is there a reference 'I shot . . .', though he frequently says a species 'was shot' by someone else, and regrets its having been killed. Bewick had made the decision not to kill whilst still a young boy. As a child he had knocked down many birds with stones, but they all escaped. One day, he scored a direct hit, a bullfinch fell from the tree, and Bewick picked it up. He said, 'It was alive, and looked me piteously in the face, and, as I thought, could it have spoken, it would have asked me why I had taken away its life. I felt greatly hurt at what I had done, and did not quit it all afternoon. I turned it over and over, admiring its plumage, its feet, its bill, and every part of it.' Bewick

then stated, quite simply, 'This was the last bird I killed.' Even when he urgently needed specimens for his book, he did not use that well-worn excuse for killing, that it was 'in the interests of science'. One respects a man whose adherence to principles is maintained even when those principles make life difficult for him.

BULLFINCH

Bewick gives us an account of his ideal bird model, a ringed guillemot, caught alive at the mouth of the Tyne, which sat quite unconcernedly under a table preening its feathers whilst being drawn by Bewick and stared at by some children. Afterwards it was taken to the shore and released. A swallow with a damaged wing was placed on his bench whilst he engraved its portrait. It attentively watched every movement he made and when he relaxed for a moment and looked at the bird, it ran up close to the graver in his hand hoping to be fed some flies. Any bird which was brought to him alive, was sure to be released again as soon as its portrait had been drawn.

The last facet of Bewick's work which remains to be discussed is his water-colour drawing. We are very fortunate in having so many of his paintings of birds survive to the present day in publicly owned collections where they may be studied. Isabella, Bewick's second daughter, presented many of the original drawings and proof wood-engravings to the British Museum in 1882. There were 138 water-colours and another 91 monochrome studies in wash, pencil or pen and ink. Not all of them were studies of birds, since the gift included sketches made for Bewick's other work, including *Aesop's Fables* first published in 1818 with the help of William Harvey, William Temple and Bewick's son Robert.

In the British Museum collection are some 76 drawings of birds and 39 sketches for the tail-pieces in *British Birds,* but Bewick sometimes discarded his first study and the wood-engraved figures do not always correspond with the original drawings. The Prints & Drawings Room also has some water-colours, among them detailed drawings of feathers. A few drawings remain in the hands of private owners, with several more in the Witt collection of the Courtauld Institute. But by far the largest number are, appropriately, in Newcastle, being taken care of by the Hancock Museum staff. Miss Isabella Bewick left these to the Natural History Society of Northumberland, Durham and Newcastle upon Tyne in 1884, and the society lodged them in the Hancock Museum.

The museum was named after the same John Hancock whom Bewick had thought to be a promising naturalist in 1825. Hancock was to publish a new edition, carefully printed from cleaned blocks and not over-inked, of Bewick's *British Birds,* in 1847. He owned an extensive museum of natural history which he gave to the townspeople of Newcastle in 1884, six years before his death. The N.H.S. of Northumberland, Durham and Newcastle upon Tyne still display their books and specimens in the natural history museum in Newcastle, known as the Hancock Museum.

Mr D. Croal Thomson made a careful study of Bewick's water-colours, and published a book about them in 1930. In it he listed over 300 water-colours of birds, though many are foreign species and these, with some others, were not used in *History of British Birds.* About 240 were the basis for the wood-engravings in this book. Bewick used pale wash colours, which are delicate and charming, but, despite the application of colour, his drawings seem pale and insipid in comparison with their vigorous and life-like monochrome engraved counterparts. Bewick enjoyed using water-colours, and he had a copy of the 1805 edition of *British Birds* printed very lightly by Wingate, and then spent many winter evenings happily colouring the figures. Some of these prints are in the British Museum Prints & Drawings Room and bear the inscription, 'coloured by Thomas Bewick' at the foot of the sheet. Lastly, a handful of drawings are in the John William Pease Collection acquired in 1901 by the Public Library, Newcastle.

In August 1823 Bewick took a well-deserved holiday in Edinburgh. He called on Patrick Neill, the naturalist and bookseller who helped Audubon find an engraver, and says of his visit that he 'was much pleased with seeing the tamed birds and other curiosities which embellished

YELLOW BUNTING

his little paradise.' Among the engravers of Edinburgh whose workshops Bewick visited were Lizars, to whom he paid his respects, and 'Mr Stewart'. This was James Stewart who did the drawings for Adams' edition of Bechstein's *Cage and Chamber Birds*, and many drawings for Sir William Jardine, which were engraved by Lizars.

Two other holidays of note were taken by Bewick. One was at Buxton, Derbyshire in May 1826 and another at Scarborough in July 1827. These were attempts to relieve Bewick's rheumatic gout, from which he suffered much pain in his last years.

For many years Bewick had looked forward to retiring and living in the country again, away from the noise and bustle of the city. His wish was never granted, and he died whilst still at work, on the 8th November 1828. He was buried beneath the tower of Ovingham Church, the parish church of Cherryburn which had been the subject in the background scenes of a number of his wood-engravings. His wife, Isabella, had predeceased him by two years after they had been happily married for forty years. Their three daughters all remained unmarried and survived their parents. The son, Robert Elliot, carried on the business until his death in 1849 and continued to print from his father's blocks and to sell his books.

The most delightful description of Bewick was written by one of his contemporaries, Audubon the American artist and author. Audubon went to have tea with Bewick in 1827, when he was in Newcastle, and wrote an account of his visit for his wife Lucy, giving her a pen-portrait of Bewick every bit as vivid as his brush portraits of birds. He had found Bewick ready to greet him, 'with his night-cap on and tobacco pouch in one hand ready to open'. The night-cap was worn by Bewick to cover up a bad burn on the back of his head which he had sustained as a child. 'Bewick welcomed me,' Audubon continued, 'with a hearty shake of the hand, and took off for a moment his half-clean, cotton night-cap tinged with the smoke of the place. He was a tall, stout man, with a large head, and with eyes placed further apart than those of any man that I have ever seen; A perfect old Englishman, full of life, although seventy-four years of age, active and prompt in his labours.' Bewick showed Audubon the 'work he was at, a small vignette cut on a block of box-wood not more than three by two inches, representing a dog frightened during the night by the false appearances of men formed by curious roots and branches of trees, rocks, etc. The old gentleman and I stuck to each other; he talked of my drawings and I of his woodcuts, till we liked each other very much. Now and then he would take off his cotton cap, and draw up his grey worsted stockings to his nether clothes; but whenever our conversation became animated, the replaced cap was left sticking as if by magic, to the hind part of his head, the neglected hose resumed their downward tendency, his fine eyes sparkled, and he delivered his sentiments with a freedom and vivacity which afforded me great pleasure . . . He was purely a son of Nature to which he owed nearly all that characterized him as an artist and a man.'

*Books illustrated by Bewick*

BEWICK, T. A History of British Birds: the figures engraved on wood by Thomas Bewick vol. I. Land birds (with text by Ralph Beilby); vol. II. Water birds (with text by T. Bewick) 1797-1804.

Wood-engravings: 208 in vol. I. (119 birds + tail-pieces); 240 in vol. II. (114 birds + tailpieces)

Artist and engraver: T. Bewick

2nd ed 1802 2 vols

3rd ed 1809 2 vols (224 figs in vol. I.; 229 figs. in vol. II)

4th ed 1816

5th ed 1821 2 vols Part I 43 wood-engravings (21 ornith.); Part II 40 wood-engravings (21 ornith.)

Supplement 1821 vol. I. 20 land birds; vol. II. 18 water birds

6th ed 1825 2 vols. vol. I. 156 figs; vol. II. 146 figs

Together, 157 figs birds + 145 tail-pieces

7th ed 1832 vol. I. 332 figs; vol. II 307 figs. (including 14 extra engravings of foreign birds)

8th ed edited by John Hancock 1847 vol. I. 325 figs; vol. II 321 figs + 20 extra tail-pieces + 14 extra figs. of foreign birds in vol. II. (Last edn printed from original wood blocks.)

# *Thomas Campbell Eyton 1809–1880*

'Eyton-upon-the-wild-moors, a parish and small village is pleasantly situated in a retired locality, two miles and a half north from Wellington . . . Eyton Hall is a handsome stuccoed residence with a piazza in front supported by twelve pillars; and in front of the hall is a small park, stocked with deer. It is the seat and property of Thomas Eyton Esq., who is the principal landowner and lord of the manor.' This was where Thomas Campbell Eyton was born in 1809 and where he died in 1880. Eyton Hall was the property which he inherited as the 23rd heir in direct male descent in 1855 when his father Thomas died. His mother was Mary, daughter of Major General Donald Campbell.

Thomas Campbell Eyton was a very energetic man of many parts. He was an active magistrate and Deputy Lieutenant of the county of Shropshire. He instituted the herd book of Hereford cattle in 1842 and continued its publication until 1860. He used his own yacht and finance to investigate the oyster fisheries around the British Isles and on returning home he prepared a report which was published in 1858 under the title *A History of the Oyster and the Oyster fisheries*. It was illustrated with some fine lithographs of the oyster and its anatomy which he had drawn from his own dissections. His ancestors were members of the firm Rocke, Eyton & Co. of the Old Bank in Wellington, and Eyton followed in this tradition, becoming a very wealthy man, though he was also a very generous one. He was a pioneer of the volunteer movement in Shropshire in 1859, becoming Captain of the Wellington Company of Shropshire Rifle Volunteers. The band of the 1st Company of Volunteer Rifles were given their instruments by him. In 1973 his descendants presented a plaque to the K.S.L.I. Museum which Eyton had been given by the Volunteers to show their appreciation of his gift of instruments to their band.

Eyton was a keen sportsman, an excellent shot and collected many of the birds which he prepared as specimens, keeping them either as skins, or mounted, or as skeletons. He hunted with the local foxhounds for many seasons. Among the numerous catalogues which he compiled, was a detailed record of the published literature on foxhunting and fishing. In addition, he encouraged local sporting and natural history societies, and took his turn serving on their committees.

When he was twenty-seven, Eyton published his first bird book. He wrote about the *Rarer British Birds* and illustrated the slim volume with wood engravings. His decision to use this type of illustration was influenced by his admiration for Bewick, and the fact that a wood engraver called Mark was at work in Wellington. Eyton lived in Wellington for some years before he inherited Eyton Hall, first at a house called Donnerville in 1842 and then in a Georgian mansion called The Vineyard in 1851.

The format and make-up of *Rarer British Birds* was

very similar to that of Bewick's *History of British Birds* which it was intended to supplement. A wood-engraving of the species was placed at the head of a page, and notes about the bird's physical appearance, where it had been located, its food, and other details of its life, followed. At the end of the letterpress a vignette was printed.

How Eyton became acquainted with Mark is unknown. In the Preface, all Eyton says about him is 'The cuts are all executed by Mr Mark, engraver on wood, Wellington.' The bird figures were beautifully drawn and executed, though rather less life-like than those of Bewick. The backgrounds lacked the same detail and skill, but were of similar subject content and scenery. The engraving was not so fine-lined, all the line work being a great deal thicker than Bewick's in all parts of the block. Instead of the series of short graver strokes which Bewick used, Mark engraved a series of ovals of quite thick lines. This very distinctive manner of engraving is most evident in the trees and stones and other foreground details.

HOODED MERGANSER

Only a few of the thirty-eight vignettes showed birds. One was printed as a tail-piece to the section on the Alpine accentor. It depicted a fox about to spring on a group of three pheasants feeding in a glade. Eyton refers to the bird figures in the Preface where he stated 'with those exceptions noted in the letterpress accompanying them, the drawings have been made from specimens in the author's collection.' They were most probably drawn by Eyton himself. In a letter to a friend dated Jan. 12, 1838, Eyton said he had been 'lately drawing a female species of common shag', and there is further evidence that he did his own drawings.

GULL-BILLED TERN

In *Rarer British Birds* forty-two species were regarded as rare, some of which one would expect, but the inclusion of others is rather surprising. White's thrush, harlequin ducks, white-winged (now two-barred) crossbills and the Virginian (now yellow-billed) cuckoo were rarities, and still are in varying degrees. On the other hand, wood pigeons and 'brown-headed gulls' (Eyton's much more sensible description of our black-headed gulls) are all too abundant today and one can hardly credit they were so rare in 1836. More understandable is the inclusion of firecrests, pomarine and Arctic skuas, perhaps also Arctic terns. Three of the best illustrations are the Egyptian vulture on the title-page, the gull-billed tern and the shorelark.

EGYPTIAN VULTURE

This publication made Eyton known to many other ornithologists. He soon started corresponding with A. R. Wallace, Charles Darwin, Professor Richard Owen and Louis Agassiz. He was a close friend of Darwin for many years, even though he strongly disapproved of Darwin's theory of evolution and was very chagrined when Darwin used some of his researches on pigeons to substantiate that theory. Another friend and correspondent was William Yarrell who was also Eyton's newsagent and supplied him with copies of the *The Times*. On 9th May 1840 Eyton began a letter to Yarrell, 'I enclose a check for the amount of my newspaper account and take the opportunity of having a little zoological chat.'

Eyton's letters to Yarrell, dated 1833 to 1840, give a vivid impression of Eyton's personality. His thoughts followed one another in quick succession—so rapidly in fact that he had no time to bother about such minutiae as full-stops, commas or capital letters, and even his grammar is occasionally suspect. His hand-writing is large and scrawling, with two or three sentences consuming a whole page. He wrote exactly as he must have talked to his friends, and he comes across as a cheerful, enthusiastic, intelligent man of many interests and boundless energy. As an example, when writing about his barnacle geese he said to Yarrell, who was a keen egg-collector, 'I trouble you with these few lines to say that my Barnacle geese have began to lay, but not in a nest, dropping their eggs in the water and by the side of it the magpies consequently come in for a share I have however procured one perfect one and now that we are on the look out shall no doubt find more.'

Two years after his *Rarer British Birds* was published, Eyton issued a *Monograph of the Anatidae* with some

SHORE LARK

wood-engravings and 24 lithographed plates. Six of the plates were coloured by hand, and were designed for Eyton by Edward Lear. One of these plates was signed 'E. Lear del.T.C.E.lith.' showing that Eyton had copied Lear's drawing onto the stone using lithographic chalk, ready for Charles Hullmandel to take a print. The coloured plates were full-page bird portraits with the background and foreground hand-coloured as well as the bird figures. The uncoloured plates showed anatomical details of intestines and the soft parts of birds. Most probably, some of these were also executed by Eyton, and others were done by Georg Scharf, a lithographer from Mainburg who had settled in London in 1816 and practised there as a lithographer. Eyton wrote to Yarrell In July 1837 asking him to allow 'Scharf to draw from your collection the trachea of teal, wild duck, longtailed duck and Egyptian goose'. At the same time he sent Yarrell some sketches for vignettes and gave Yarrell permission to use whatever he thought suitable for his

FIRE-CRESTED WREN

book then being issued in parts. The wood-engraved figures in the *Monograph of the Anatidae* bore no signature and it is not easy to tell from the slight drawing required for a leg and foot of a bird who had done the drawing or engraving. It is possible that Mark did these engravings, or even Eyton himself, since he was very versatile. Jardine sometimes drew on the wooden block for an engraver to grave, so Eyton might have used this method.

Eyton named quite a number of ducks and described them for the first time in this monograph. Apart from those which were new to science in 1838, there was another one which is now known either as Eyton's tree duck or the plumed whistling duck. The scientific name is *Dendrocygna eytoni,* and Eyton explained that 'This beautiful and unique bird, which Mr Gould has done us the honour to call by our name, is in the possession of the Zoological Society, and was received from Australia.' This was one of his sources for these rarer ducks. Other sources included the specimens owned by the Earl of Derby, Darwin, Yarrell and 'the author of Birds of Europe' i.e. John Gould. Eyton also used the 'South African species brought home by Dr Andrew Smith now exhibited in Egyptian Hall' in Piccadilly. Dr Andrew Smith was Jardine's collector and also collected skins on his own account. He was an early explorer of the interior of South Africa and returned laden with objects of natural history. He sold off the items at the end of his exhibition in 1838.

WHITE'S THRUSH

Eyton was the recipient of drawings and specimens from other naturalists and professional collectors. Some of these worked abroad, others in England. On the 3rd April 1834 he told Yarrell, 'I have taught a man on the Welsh coast to make sketches and he really is a very good hand at it but as he makes many more than I want if you could find anybody who wished to be supplied with sketches of those kinds on that coast of course much below the London price as the birds are procured for a mere nothing comparatively speaking he will be happy to supply them and I shall be much obliged to you.' By the time he inherited Eyton Hall he had amassed a large collection of both drawings and specimens. In 1855 he moved to the Hall and built a museum there to house his collections. He was particularly interested in anatomy and his museum was rich in skeletal preparations. His enthusiasm for this branch of ornithology was conveyed in a letter to Yarrell in 1838, 'I have every old pitcher about the place full of bones and solutions my room garnished with skins and a long series of jars full of organs of digestion of which I am preserving as (?) I can get in salt and water which is cheap enough and seasons well.' The same year in which he built his museum, the Zoological Society Museum was dispersed and provided Eyton with a golden opportunity to procure some rare species. He bought a number of foreign birds, many of them brought to England by the *Sulphur, Beagle* and other expeditions, which returned with some original or type specimens never described before.

BUFF-BREASTED SANDPIPER

Three years later he published a list of the specimens in his museum. *The Catalogue of Species of birds in his possession,* 1858, ran to 367 pages, though it was a small book in size. The discoveries he made during the course of many years' study of anatomy of birds, based on these specimens, were disclosed in the last book he published, *Osteologia avium,* between 1871 and 1878. This was a voluminous work on the skeletons of birds and occupied Eyton to the end of his life. A supplement was issued in 1878, just two years before he died, and it included further examples of Scharf's work. Eyton died in October 1880 after a long and tedious illness.

Eyton had married Elizabeth Frances, the daughter and coheiress of Robert Aglionby Slaney, M.P. for Shrewsbury, and they had seven children. His eldest son Thomas Slaney Eyton decided to sell the museum. Dr R. B. Sharpe made a selection of the type specimens for the British Museum but said that finding them had been difficult because the labelling on the specimens in Eyton's scrawling handwriting was illegible. Though Dr Sharpe missed a few of the type specimens he purchased 108 skeletons and 205 skins in 1881, whilst E. Gerrard purchased the remainder of the collection.

Unfortunately, Eyton Hall has now passed out of the ownership of the Eyton family, and the manuscripts, drawings and other papers left by T. C. Eyton have vanished. It is possible, however, to obtain his *Rarer British Birds* on the second-hand book market quite cheaply, though the duck monograph is neither easily nor so reasonably bought.

*Eyton's books with illustrations*

EYTON, T. C. A History of the rarer British birds . . . discovered since the time of Bewick. 1836 80 wood-engravings (42 figs birds) [The first issue had the fig. of rock grouse omitted and a corrected leaf was distributed.]
Artist: Eyton
Xylographer: Mark

EYTON, T. C. A Monograph of the Anatidae or duck tribe 1838 24 plates (6 col. lithogs) and wood-engraved diagrams.
Artists: E. Lear 6; G. Scharf; T. C. Eyton (anatomical)
Xylographer: ?
Lithographers: C. Hullmandel; T. C. Eyton 1 signed pl.; G. Scharf.

# *William Yarrell 1784–1856*

William Yarrell was a paragon of virtue, and more remarkable, an attractive one. He was even-tempered, generous, patient, thorough and modest. He was a successful businessman and a popular author, yet his way of life remained simple. In short, it is difficult to fault Yarrell, and the more one learns about him, the less one is inclined to be critical.

He is important as the ornithologist who provided the standard illustrated textbook on British birds in the 19th century. Just as Ray & Willughby, and Pennant stand head and shoulders above their contemporaries in this respect, so does Yarrell in his century. Yet another great stride was made in recording the British avifauna and the way in which Yarrell presented the information became a model for future authors.

Yarrell was born on the third of June 1784 in Duke Street, St James's, where his father and his uncle were in partnership in a newsagent and book-selling business. He went to school in Ealing with his cousin Edward Jones, the pair of them later running the business established by their fathers. His cousin died in 1850, leaving the business to Yarrell. However, though it provided the background to his life, and was run well enough to provide him with a good income, Yarrell's interests were less in the business and more in the naturalist's world. He was an enthusiastic sportsman who rode and shot well, and was also a keen angler. He often visited Cambridgeshire and Hertfordshire where he shared the shooting on various estates with a friend. His mother (Sarah Blane before marriage) was a native of the beautiful Hertfordshire village of Bayford, and he was familiar with the birds of that neighbourhood from an early age. His frequent visits to the countryside gave him some knowledge of common birds, then in 1823 he began to keep records of the appearance of rarer species.

In 1817 Yarrell became a member of the Royal Institution and attended lectures in chemistry and natural history, but soon abandoned the chemistry lectures. He studied hard, and gradually became extremely knowledgeable in a number of branches of natural history, though he found fishes and birds to be the most fascinating subjects. By 1825 he was ready to join the select circle of naturalists in the city, and made his debut by publishing a careful study 'On the occurrence of some rare British birds' in the second volume of the *Zoological Journal*. After this he quickly became known and accepted by other scientists. In the same year, he started writing to Bewick, now an old man and at the end of his career. Yarrell admired Bewick greatly, and helped him with his last edition of *A History of British Birds* by sending him some specimens of the rarer species. Bewick engraved the figures from these specimens and acknowledged Yarrell's assistance.

In the *Transactions* of the Linnean Society in 1830,

Yarrell described the difference between the whooper swan and another very similar swan which he had first identified as a British bird in 1824. To the new swan he gave the name *Cygnus bewickii,* 'Thus devoting it to the memory of one whose beautiful and animated delineations of subjects in natural history entitle him to this tribute.'

In September 1828, shortly before his death, Bewick visited Yarrell in London. Bewick was then seventy-four and had made this journey from Newcastle against his will and better judgment in order to sell off his 'great remaining stock of Birds and Quads.', as he referred to his two books. In a letter to Dovaston, Bewick narrated, 'I

STARLINGS

called upon Mr Leadbitter Bird preserver Ec. to the British Museum—& then upon Mr Yarrel from them I met with a most kind reception & offers to do everything in their power to serve me in my ornithological or on my intendig fishy publication—I never in my life was so gratified at seeing any thing, as I was with Yarrell's Museum—he had left nothing untouched that could assist him in probing ornithological knowledge to the bottom—he invited me most pressingly and kindly to dine with him, chiefly for the porpose of my meeting with Mr Vigors there—but I dclined I had made up my mind *as I thought,* not to dine with anyone for fear of being *Buzzed to death in London . . .*' This must have been an extraordinary meeting, with Yarrell a little in awe and trying his hardest to put at ease a Bewick who was agitated and who hated London, yet was consumed with curiosity whenever a museum or anyone who could talk about fishes and birds came within his reach. Yarrell succeeded in overcoming Bewick's apprehension by his kindliness and understanding. He must have been saddened on hearing of Bewick's death only two months

HAWFINCH NEST

HAWFINCH

later, but would also have been pleased that he had met Bewick and given the old engraver a most enjoyable few hours among the fishes and birds in his collection.

Yarrell also became acquainted with Selby, with whom he stayed on some occasions at Twizell House, and with Sir William Jardine, John Gould, N. A. Vigors, and many others. Some of them he met at the Zoological Society meetings, for he joined the Society on its formation in 1826 and later served as Secretary and Vice-President. He was elected a Fellow of the Linnean Society in 1825 and served that society as its treasurer from 1849 until his death. His interest in insects was shared with other founder-members of the Entomological Society from 1833 onwards and he was treasurer from 1834 to 1852. Yarrell was very hospitable and fond of society so that his house in Ryder Street was well-known to his fellow naturalists. Leonard Jenyns, in his memorial of Yarrell, gave one of the reasons he was so popular. 'One marked feature in his character was his readiness to make himself useful to others, and especially to assist in their researches, if appealed to, all interested in those branches of Natural History, to which he was himself attached.'

BLACK-TAILED GODWIT

Apart from his own newsagent business and the paperwork involved for all the societies to which he belonged, he still found time to write many articles for the journals and periodicals which were multiplying in number in the mid-19th century. The societies issued their own publications, to which Yarrell contributed a number of papers, e.g. the *Transactions*, *Proceedings* and *Journal* of the Linnean Society; the *Transactions* and *Proceedings* of the Zoological Society; the *Entomological Magazine*. His papers and articles may also be found in the *Zoological Journal* and *Philosophical Magazine*, and the two commercial periodicals *Annals & Magazine of Natural History* and the *Zoologist*. Yarrell was a man of unbounded energy and was fortunate enough to enjoy good health up to the last two years of his life.

After many years of careful preparation, collecting and checking the records of the appearance of birds in the British Isles, Yarrell was at last ready to start issuing his book about British birds. His *History of British Birds* was issued in parts, the first one appearing in July 1837. Subsequent parts appeared at bi-monthly intervals until the work was completed in May 1843. The book was produced on the same plan as Bewick's and they shared the same title. Bewick's influence did not end there, for Yarrell thought so highly of the excellent wood-engravings in Bewick's books, that he determined to illustrate his own work with engravings. In this first edition of his *British Birds* the three volumes had 579 figures. Various supplements and new editions followed in response to public demand and in order to keep the work up to date.

Alexander Fussell's drawings of the birds, though closely imitating Bewick's in style and composition, were stiffer and less life-like than those of the old master. The rural background scenes, on the other hand, had just as much charm and authenticity as Bewick's. Alexander Fussell was about twenty-three when he began to draw for Yarrell. His father was Joseph Fussell, an artist who taught oil painting to students in Birmingham before he moved to London, in 1821, and started showing his landscape and religious pictures in London exhibitions. Between 1821 and 1845 he exhibited fifty paintings. Alexander was the eldest of Joseph's three sons who became artists, and was born *c.* 1814 at Birmingham, so he would be six or seven years old when taken to London, where he was most probably taught to paint by his father. Alexander made his debut as an artist when he exhibited his first picture at the British Institution in 1838, following up this success a year later when another of his pictures was accepted by the Royal Academy. He worked steadily at his painting for many years, exhibiting either landscapes or seascapes until 1881, the one exception being a portrait of his wife and baby daughter Elizabeth in 1847, again at the Royal Academy. He married Ann Booth, the daughter of a Runcorn (Cheshire) merchant, in 1840. By 1847 they had settled at no. 2 Oakley Square, Camden Town, quite close to the Zoological Gardens in Regent's Park which Alexander frequently visited in order to draw some of the rarer British birds kept alive in captivity there. In 1864 he moved still nearer to the gardens, but in 1875 he took Wavertree Cottage in Sidcup and thereafter lived out of London. When he died on May day 1909, aged ninety-five years, he was attended by another daughter called Ellen, at his home in Worthing, Sussex.

The Fussell family would have been so well known in London's art circles in the 1830s and 1840s, that Yarrell saw no need to explain who his artist was, nor to give any further information about him. Alexander Fussell worked on the first three editions of Yarrell's *History of British Birds* from 1836, or earlier, to 1856, but all Yarrell said of the engravings, in his preface, was 'To Mr Alexander Fussell, for the ability, invention and good taste which have enabled him to give truth of character, variety and effect to nearly five hundred of the drawings on wood here employed, my best acknowledgements are due; and more particularly so to Mr John Thompson and his sons, for the skill, the zeal, the success, and I may add the pleasure, with which they have laboured throughout this very long series of engravings.' Though this indicated a clear division of labour between Fussell the artist and the Thompsons as engravers, some of the wood-engravings e.g. the lesser and mealy redpolls, twite, goldfinch, siskin, were signed 'Thompson del et sc.' indicating that Thompson was the artist as well as the engraver.

MEALY REDPOLL

The wood engravings were done by John Thompson, a former pupil of Allen Robert Branston, and two of his sons. John Thompson (1785–1866) was one of the most distinguished wood-engravers of his time and in 1855 won a *médaille d'honneur* at the Paris exhibition. He also exhibited his engravings in Britain, as well as some landscape paintings. These were shown at the Suffolk Street Galleries belonging to the Royal Society of British Artists, between the years 1824 and 1832. Much of his work was done for book illustrations, over 900 of these were engraved by him from John Thurston's designs and published by the Chiswick Press. As a compliment to Thurston, John Thompson's elder son by his wife Harriott, was called Charles Thurston Thompson. This son was also a very good wood-engraver whose work was published by Longman and Van Voorst. He later became interested in photography and was appointed as

COMMON PARTRIDGE

the official photographer to the South Kensington (now Victoria & Albert) Museum in 1857, a year after Yarrell's death. John Thompson's second son, Richard Anthony, also abandoned wood engraving when he was appointed to a post at the South Kensington Museum in 1851. He quickly showed outstanding administrative abilities and eventually held the post of museum superintendent. Richard Anthony would only be about seventeen years old when his father accepted the commission from Yarrell for the engravings in *British Birds* and his contribution was probably quite small. All the Thompson's engravings were delicate and charming as well as being accurate representations of the birds.

The engravings were signed only infrequently by Fussell and Thompson, many of them being left unsigned by either of them. The Thompsons' signatures varied considerably but whether the differing signatures were indicative of the contribution by the three Thompsons who were engravers at this time, is not certain. Some of the variations are, 'JT del et sc.'; 'Thompson del et sc'; J. Thompson del et sc'; 'J. Thompson sc'; or just 'Thompson'.

In addition to the bird figures in *British Birds* Fussell drew many of the tail-pieces following the example set by Bewick and Eyton. However, Yarrell improved on the subjects of the tail-piece vignettes he commissioned from Fussell. Whereas those of Bewick and Eyton were not often ornithological, Yarrell's were apposite vignettes depicting anatomical details such as the breastbones and trachea, the asymmetrical ear-positions of owls, scenes of falconry and wildfowling. Some of them are charming, particularly when Fussell and Thompson showed the nest and eggs of the species, or a young chick.

DUNLIN
CHICK

Another kind of 'tale-piece' as Bewick would have called it, was the vignette showing a man and woman chatting in a wood at the end of the section on the woodchat shrike!

The sources which Yarrell used were as varied as one would expect of a man with such a wide circle of friends. He constantly makes reference to other people's collections of specimens, for example, in the extensive museum of Sir William Jardine he saw the skin of a wood sandpiper. He cultivated the acquaintance of agents who collected and preserved birds, and taxidermists who kept an eye open for unusual skins. Yarrell got a rock thrush from Joseph Trigg by this means. Trigg had shot it whilst it was sitting in an ash tree at Therfield near Royston on 19th May 1843 and taken it into the town to be stuffed by John Newman of Royston. Yarrell added this species to the British list when he described this specimen. Added to these sources, Yarrell had a considerable collection of his own which included a series of British birds and their eggs.

GOLDFINCH NEST

The text of his *History of British Birds* was written in a clear, simple style. The information given was based on sound research and scholarship. Yarrell had spent months sifting evidence of the appearance and locations of birds, and sorting out the multiplicity of local names which the species had acquired over the years due to insufficiently detailed descriptions of them. He also noted the habitats and diets of the birds and their nesting routines. His knowledge of the birds' anatomy was profound but he kept this aspect of the subject in proportion with all the others, and trod warily when he was in disagreement over the classification of the species as a result of his anatomical studies.

TURNSTONE

The charm of the engravings and their accurate delineations added to the appeal of *British birds* which soon became an extremely popular textbook. It was stated by his publisher and friend, Van Voorst, that with this book and his *History of British fishes*, published in 1836, Yarrell made £4,000.

Alexander Fussell and the Thompsons appear to have been entirely responsible for the illustrations in the first three editions which Yarrell saw through the press before his death. The book was still so much in demand, that a fourth, completely revised and virtually re-written edition was published in 1871-85 by Professor Alfred Newton and Howard Saunders, with 564 wood engravings in four volumes. In this last edition there were additional illustrations by the artists J. G. Keulemans, Edward Neale and Charles Whymper, and the extra engravers were R. C. West, Madame G. Pearson and the Dalziel Brothers. Keulemans and Neale are better known as artists who worked for the lithographic medium, though they also did many drawings for engravers to copy. The text reveals that Keulemans drew many more

WHITE-WINGED LARK

of the birds than the two which bear his signature (the tawny pipit and white-winged lark). Edward Neale signed eleven of the bird figures (8 in vol. 1 : 3 in vol. 3).

PALLAS'S SAND-GROUSE

Until recently, Edward Neale was just another of the many Victorian artists whose name was known only to a few bird book collectors. This is a pity, because his bird drawings were good likenesses and most sensitively executed. Unlike Fussell and the Thompson brothers, Neale did not come from a family of artists, but was born in the village of Speenhamland, Berkshire, and baptised in May 1833 the son of John Neale, a grocer, and his wife Mary Ann. He not only made many beautiful small drawings for bird book illustrations, but exhibited some of his large bird paintings at the British Institution, Royal Academy and Suffolk Street Gallery between 1858 and 1880. He was a bachelor, living in London most of his life. He died in Middlesex Hospital on 11th November 1904, leaving his pictures and sketches to his two remaining sisters out of a large family.

The 4th edition of Yarrell was as good as the first edition with regard to the quality of the engravings. The blocks for these engravings were owned by the publishers Gurney & Jackson. When Howard Saunders needed some illustrations for his book *An Illustrated manual of British birds* in 1889, he decided to make use of them. This had a definite advantage which Saunders pointed out to his readers in his preface. 'The low price is due to the fact that the publishers own blocks from which the illustrations were struck for the four editions of Yarrell's British birds though to these have been added woodcuts of many recent wanderers to Great Britain . . . while fresh engravings have taken the place of unsatisfactory originals of the Marsh-Harrier, Gos-Hawk, Merlin and Great Auk.' Once again, the ability of good quality wood blocks to stand up to repeated use was demonstrated.

ISABELLINE WHEATEAR

Only eleven more figures were added for the second edition of Saunders' book, in 1898–9, George Lodge, a meticulous artist, supplied illustrations of new birds added to the British list and he also did new illustrations of twenty other species. Lodge was an engraver as well as artist and some of his illustrations in Saunders' work were signed 'G. E. Lodge del et sc.' Saunders had edited the last two volumes of the 4th edition of Yarrell. His own *Illustrated manual* was, in a sense, still keeping Yarrell's book up to date and alive, for the *Illustrated manual* included the whole essence of Yarrell's text as well as his engravings.

SOCIABLE PLOVER

Yarrell had been fortunate to suffer no illness and he was still actively interested in pursuing his hobbies when he was seventy. Then he had a severe attack of fever, but he recovered. After this he had a slight paralysis, but regained his faculties and strength once again. At the end of August 1856 he went by sea to Great Yarmouth for a

HARLEQUIN DUCK

rest and to enjoy the sea-air. The day after he arrived, he died. His body was taken to Bayford and buried in the churchyard where other members of his family lay. Their headstones are now grouped together. He left property worth £17,000 and his extensive library and valuable collection of British fishes and birds were auctioned.

In St James's Church, Piccadilly, there is a marble memorial tablet in the south aisle. It was erected to his memory and has a medallion portrait, supported by two swans. This is a fine tribute to the man who was inspired by Bewick and named the swans after his friend.

It would not be fitting, however, to end this account on a sombre note. Yarrell was greatly appreciated by his friends as a good, cheerful companion. Jenyns was one of his friends and he has left us a picture of Yarrell the host which tells us much about this convivial bachelor. 'Yarrell was particular about his dinners and cookery, and had everything very nicely served up. He was an excellent carver, and his knowledge of comparative anatomy told him where exactly to plant his knife in dissecting for the table, fish, game or any small joint that required nicety in helping.' From Sir John Richardson's account of his life, Yarrell emerges as a good singer of glees, and a frequenter of theatres in the front row of the pit. He memorised the popular songs of the day and sang them with great gusto. He once silenced the buzz of conversation at a large evening party at John Gould's house when he burst out singing in a stentorian voice, 'I'm afloat, I'm afloat', the opening words of an amusing sailor's song. He hardly ever missed a dinner of the Linnean Club and when he joined in the society's country excursions he was one of the liveliest members present.

A truly attractive personality.

HOOPOE

*Yarrell's artists and xylographers*

YARRELL, W. A history of British birds 1837–1843
3 vols 520 bird figures + 59 tail-pieces = 579 wood-engravings
Supplement 1845 14 wood-engravings
Supplement 1856 18 wood-engravings
2nd ed 1845 3 vols 535 wood-engravings
3rd ed 1856 3 vols 550 wood-engravings
Artists: Alexander Fussell; J. Thompson
Engravers: J. Thompson; C. T. Thompson; R. A. Thompson
4th ed by Alfred Newton and Howard Saunders 1871-1885 4 vols 564 wood-engravings
Artists (number clearly signed): Alexander Fussell 60; (J) Thompson 27; Edward Neale 11; J. G. Keulemans 2; Charles Whymper 3
Engravers (no. clearly signed): the Thompsons 27; R. C. West 1; Mme G. Pearson 2; Dalziel Bros 4

# *William Dickes 1815–1892*

William Dickes was by nature an artist, by profession a master of printing crafts. He undertook work 'at moderate charges', including, 'drawing and engraving on wood, steel engraving, lithography, chromolithography, letterpress colour printing, and copper plate engraving' according to an advertisement in the 1860s. Dickes was a most painstaking worker, nevertheless he produced a vast quantity of prints, as cheaply as possible, whilst maintaining the highest standards. In his prime, he was employed as the principal colour printer by the S.P.C.K. and the Religious Tract Society among a large number of publishing houses, and by many publishers of periodicals including *People's Magazine, Gentleman's Journal,* and *Queen.*

Dickes was born on 7th May 1815 at Beechencliff near Bath. He was taken to London when young, and later apprenticed to Robert Branston the younger. In 1835 he decided to improve his drawing and enrolled as a student at the Royal Academy Schools where he won silver medals for some of his drawings. Lizars knew of the work of young Dickes (perhaps through Branston to whom he was related) and about 1840 he asked Dickes to provide some water-colour drawings of animals for Sir William Jardine's Naturalist's Library series. A small number of Dickes' drawings were engraved on steel by Lizars to illustrate the pocket-sized volumes. This gave Dickes his first experience in drawing natural history subjects for reproduction as prints to illustrate books.

In 1841 Dickes married Sarah Bloomfield and their first son William Frederick was born in 1843. William, and a second son Walter, were later to be taken into their father's business, which was renamed William Dickes & Co. after 1864. Altogether, Sarah and William had nine children, six sons and three daughters.

The business had been established *c.* 1840 and flourished until 1878. In 1842 and for the following five years, Dickes was employed in making drawings for the Abbotsford edition of the Waverley novels, and he also engraved a proportion of the wood-engravings. In addition, Robert Cadell commissioned him to superintend the other wood-engraving necessary to illustrate this edition. Many artists and engravers were engaged on the same work and Dickes thus came into contact with the main group of London print-makers so that other work quickly came his way. He was a fine draughtsman and very good engraver and many of his wood-engravings in monochrome are to be found in Victorian books of the late 1840s up to the 1860s.

Dickes was only one of a number of contemporary skilled engravers who produced hundreds of small figures of birds for the numerous books on natural history aimed at educating and entertaining the middle class in the last half of the 19th century. They may be obscure and forgotten now, but these small treasures give great delight when they are rediscovered today in secondhand book shops having lost none of their power to charm.

MALLARD

MOORHEN

Dickes also attempted colour printing from wood blocks, largely as a result of his being dissatisfied with his early attempts at chromolithography, or printing in colour from lithographic stones. The chromolithographs were incapable of reproducing fine detail for the work he was commissioned to do on reward cards and similar small prints. So about 1840 he took out a licence to enable him to use Baxter's method of printing in colour from wood blocks.

Dickes had also been in contact with Robert Branston junior about the time he had been experimenting with metal plates to print in colour in relief. A number of printers were making an effort to print in colour and so eliminate the time-consuming hand-painting of prints. Dickes made sufficient progress with his attempts to have a stand, like Baxter, at the Great Exhibition of 1851. Their colour prints were prepared from a key aquatint plate which printed the outline of the whole picture, and engraved wood blocks, one for each colour. Sometimes as many as ten wood-engravings were used to apply the tints.

PTARMIGAN

The earliest of Dickes' ornithological prints appeared in the year of the Great Exhibition. They were in a book called '*Eggs of British Birds* by C—— J——' and had a picture on the title-page as well as 7 plates. C. A. Johns was probably the author since this title was followed in 1854 by *Birds' Nests* by him, published by the S.P.C.K. and illustrated by Dickes using the Baxter process. Also in

1854, Dickes produced some pictures of poultry which gave the author Martin Doyle a great deal of satisfaction. Breeding domestic poultry was fast becoming a popular hobby about this time, and Doyle employed Charles H. Weigall (a painter and exhibitor of domestic subjects in the London galleries over a long period) to draw the live birds and then asked Dickes to make 18 prints from these drawings to illustrate the text of his *Illustrated Book of Domestic Poultry*. Dickes engraved and printed them, some by the Baxter process and some from his own early copper chromograph plates which were a slight variation on the Baxter method. Doyle declared in the Preface, 'If my shadings indeed could equal the elegant tracery of Mr Dickes in the pictorial department, I should be quite satisfied.' This was a flowery Victorian way of drawing attention to the skill and apparent facility with which Dickes could engrave flowing fine lines and include great detail in the feathering of the birds. His colours had great clarity in many prints, though they could occasionally be marred by garish tints. He used so many blocks and different colours on one print that it was easy to spoil the effect by using even one slightly wrong shade or tint.

WATERRAIL

Dickes was employed for many years as colour printer to the Society for Promoting Christian Knowledge. The

1. FULMAR PETREL. 3. STORMY PETREL.
2. MANX SHEARWATER.

Society was founded in 1698 and besides their evangelising work, carried on a large publishing business. In 1887-8, for example, no fewer than 5258 bound books were issued by them, apart from Bibles and Prayer Books, tracts, etc, and their profits were £6,600. This Society published many natural history titles of a popular nature which were cheap enough to be bought by almost anyone who could read. Dickes was involved with a considerable number of these titles, some of them by named authors, whilst others remained anonymous.

C. A. Johns was the author of *Birds of the Wood and Field,* published by S.P.C.K. in 1859 and 1862 with 12 of Dickes' colour prints. The Reverend Charles Alexander Johns was a graduate of Trinity College, Dublin, where he was ordained in 1841 and became a Head Master of Helston where he lived until 1863. Whilst at Helston he wrote the bulk of his natural history volumes. He had less time to devote to writing after he opened a private school in Winchester and founded the Winchester Literary and Scientific Society. He saw his titles reprinted many times before his death in 1874 and some were still being reissued in response to demand many years later in the present century.

OSPREY

The best known ornithological title which Dickes illustrated was another S.P.C.K. publication, Anne Pratt's *Our Native Songsters.* This was published in 1852 with a frontispiece 'Milking Time' and 72 plates illustrating British song birds. The name of each bird appeared under the print. The plates were colour printed from a number of blocks and the engravings were beautifully executed. Each picture had a pale background with a few trees in light green tints, and the bird was placed in the foreground and engraved with delicate detail. The prints were so attractive that a set of 12 of them was issued separately, with the name of the bird embossed at the base of the picture, for use as record cards. Anne Pratt was primarily a botanical illustrator, and Dickes engraved many of her drawings of ferns, grasses, sedges and flowering plants, but her birds were equally charming and reasonably accurate. Anne Pratt was born in Kent and settled at Dover, where she wrote most of her books and drew the plants which had been gathered and then brought to her by her sister. Anne was an invalid and her beautiful drawings were eloquent testimony to her indomitable will and triumph over her disabilities. She married in 1866 and her husband, John Pearless, took her to live first at East Grinstead and later at Redhill. She died at Shepherd's Bush in July 1893 having given pleasure to thousands of readers of her natural history books.

BARN OWL

RING
OUZEL

REDBREAST

1. THICKKNEE. 2. GOLDEN PLOVER.
3. LAPWING. 4. TURNSTONE.

No doubt, Anne Pratt was sustained by her faith, and Dickes was also a deeply religious man. He was connected with the Religious Tract Society for many years. He did the engraving for two undated volumes (published in the late 1850s), issued anonymously by the Religious Tract Society under the overall title *British Birds* but one volume dealt with *Land Birds* and the

1. Ruff. 2. Dunlin.
3. Gray Phalarope. 4. Collared Pratingcole.

second with *Water Birds*. Each had 12 prints of full-page illustrations carrying three or four figures of birds engraved and signed by Dickes. The birds are freely and lightly drawn, some of them being very good and their backgrounds are full of atmosphere.

Dickes exhibited his work at the Paris Exhibition of 1855 and won a medal, and again at the International

Exhibition of 1862 in London. In 1860 he had invented a method of printing on a power machine whereby the old hand process of laying on successive colours was superseded. This meant that the production of prints was cheapened considerably, due to the speed with which they could now be printed. For the first time, colour printing became commercially feasible for publishers who wished to include colour prints in their magazines and books. This achievement was largely due to William Dickes' invention of the chromograph machine and it revolutionized the art of printing in colours until it was eclipsed by the adoption of a power machine to print chromolithographs in 1873. Five years later, when Dickes' business had fallen off due to the competition from chromolithography, he retired.

Since no further coloured bird prints by Dickes after 1860 have so far come to light, we need not discuss the way in which Dickes produced the chromograph prints on his power machine. This was fully described in Docker's book *The Colour Prints of William Dickes*, published in 1924 (Chapter VIII). One small point of interest for ornithologists, however, is that the firm Messrs Kronheim & Company adopted Dickes' method and did some good work by it. Their prints were made from etched copper plates, sometimes between ten and twelve being used, and it is likely that Kronheim's prints for Cassell's *Book of Birds* issued 1869–73 by the Religious Tract Society were printed by this method.

Dickes was in retirement from 1873 until his death on 29th February 1892 at Brixton. He had bought a house at Loughborough Park about 40 years before, when his business was flourishing, and retired there. A photograph of him was reproduced in Docker's book, and the short biographical notes give a few personal details. Dickes was nearly six feet tall, good-tempered and extremely industrious, and a good husband and father. He was a talented painter in both water-colours and oils and evidently was also a good musician. He could play the flute, violin and piano and had a fine tenor voice. There is much more one would like to know about William Dickes and many other artists and engravers who laboured so conscientiously and successfully at the task of informing an ever-increasingly literate mid-Victorian population.

It is not surprising that we know so little of them today, for even whilst they were still at work, they remained obscure figures. The Christmas *Bookseller* of 1864 remarked on this and placed Dickes in the contemporary publishing scene. 'It would be unfair not to award the amount of credit due to our English printer-artists who have done so much to raise the public taste for works of art. Amongst these a high place must be accorded to Mr William Dickes of Old Fish Street whose work is scarcely known except as that of any other printer who may be employed to illustrate books for publication. Yet a very large number of these pleasing little pictures issued by the Christian Knowledge Society and published under various forms are from his office and are either drawn by himself or under his superintendence.' Dickes had done the drawings for his *Book on Figure Drawing* and produced many of the original drawings for his other plates in botanical and natural history books. He was truly a disciple of Bewick in the tradition of the artist-craftsman and naturalist. In these days of specialisation and careful demarcation, we can only look back, lost in wonder and admiration, on these men who were so skilled at such a wide variety of processes and methods and master of so many artistic media.

*Some bird books illustrated by William Dickes*

Anon. British Birds: Land Birds 12 monochrome plates engraved by Dickes. no date.
British Birds: Water Birds 12 monochrome bird plates and 1 illus. of Ranworth Decoy, engraved by Dickes. n.d. but pubd post 1856.

Anon. Birds of the Seashore. 1858 S.P.C.K. 12 colour plates. 1 print, of the Gannet, signed by Dickes.

DOYLE, M. Illustrated Book of Domestic Poultry. 1854 18 col. plates
Artist: C. H. Weigall
Engraver and colour printer: W. Dickes.

FAIRFIELD, Mrs. Stories about Birds. 1859 1 col. print of Goldfinches and Nest by W. Dickes.

JOHNS, C. A. Birds of the Wood and Field. 1859–62 S.P.C.K. 12 col. plates
Engraver and colour printer: W. Dickes.

PRATT, A. Our Native Songsters. 1852 S.P.C.K. 72 col. plates of birds.
Artist: A. Pratt
Engraver and colour printer: W. Dickes.

STOCK-DOVE

# *Benjamin Fawcett 1808–1893*

Fawcett was the third eminent wood-engraver to make his name with a book about British birds. Bewick had shown that monochrome wood engraved illustrations could adequately delineate birds for both identification and decorative purposes. Baxter had demonstrated that colour printing from engraved wood blocks was attractive and the colours reasonably accurate. Benjamin Fawcett's first bird book was illustrated with hand-coloured wood-engravings. When he turned to printing in colour, he devised his own technique, using only wood blocks and ordinary inks, unlike Baxter who used both wood and metal plates and oily inks. The drawings which Fawcett copied were in water-colours, and the opaque tints of his printing inks imitated the body-colours most faithfully. Baxter's prints were more usually copies of oil-paintings. Many of the drawings for his earliest work Fawcett did himself and much of the output at the commencement of his long career was in the field of natural history. In addition, Fawcett was his own printer—an important factor in quality control—and had the parts of his books bound up in his own workshop.

This remarkably energetic man worked in Driffield, Yorkshire. He was born in Bridlington, the son of a sailor George Fawcett, and his wife Sarah. Soon after his birth, his father died in the Philippine Islands and his mother was left to bring him up alone. She gave him a good education locally and then, in 1823, apprenticed him to William Forth, a bookseller, stationer and printer in Bridlington.

We do not know where he learnt to draw, or engrave on wood. His biographer, the Reverend M. C. F. Morris (the son of F. O. Morris) has little to say about his early years. One small, but significant fact emerges from this source. Fawcett made the acquaintance of Bewick as a young man. The meeting must have been before he was twenty, and made considerable impact on Fawcett who admired Bewick for his work, his aims, and his character. In many ways he emulated the old engraver, who died three years before Fawcett set up in business on his own.

In 1831 Fawcett moved to Driffield and opened a business as a stationer and printer. He married Miss Woodmansey about this time, and they had two sons and daughters. His first premises were in Middle Street where he remained for nineteen years. His work during this period consisted mainly of cheap illustrated children's books, drawing-books, primers and copy-books. One of the copy-books had an engraving of a tiger on the cover. This was an exact reproduction of the tiger in Bewick's *Quadrupeds*, only the figure was reversed. Whilst this demonstrated his admiration for Bewick, it is also a reminder that an artist still had no effective copyright for his work. Many other Fawcett illustrations of this period show Bewick's influence, none more strongly than a thirty-six page book *Stories about birds*, published in

Leeds. The birds which Fawcett drew best for this little book with 17 bird figures were the goldfinch, raven, kingfisher, wood pigeon and goldcrest.

TAWNY OWL

Right from the commencement of his own business, Fawcett showed a remarkable ability in selecting excellent workmen as his employees. He took on a number who were to remain with him for many years and who formed part of a faithful team of workers. Their long service permitted the printing of a vast quantity of beautiful engravings during the course of the next half century. One assistant, John Stabler, looked after the preparation of the wood and planed the blocks ready for use. He was also the lead-moulder for the stereotype foundry which Fawcett established. W. D. Ridley was one of his very competent engravers.

STONECHAT

Concerning the materials with which he worked, Fawcett was most particular. Bewick had recognised the importance of this side of the engraving process. They both took pains to obtain the best boxwood for the blocks. Fawcett used boxwood imported from Turkey. On delivery the logs were immediately sawn into slabs 1″ in thickness with a circular saw worked with the foot. The bark was left on the slabs and these were arranged on wire racks and left to mature for three years at least before they were used.

The colouring, first applied by hand and later printed from wood blocks, was also very strictly regulated and supervised. Fawcett often made his own inks from the costliest and best powders and the most expensive varnishes. His aim was to reproduce the tints and shades of the water-colour drawing exactly. He supervised this part of the work personally, each morning inspecting the colour room and looking at the work of each of the girls there. When the business increased, this room was placed under the supervision of Miss Augusta Woodmansey, a relation of his wife. After the death of his wife, about 1841, Fawcett married again. His second wife was one of the colour girls, Martha Porter. She, and their four daughters (they also had six sons) did much of the colouring. Although he printed in colours from 1852 onwards, a close inspection of those prints shows evidence of hand-touching having been used in nearly every case to finish the print to Fawcett's complete satisfaction.

In 1844 when Fawcett was well-established, and had a flourishing business, he met the Reverend Francis Orpen Morris who was appointed Vicar of Nafferton, a small village about two miles from Driffield. They were about the same age and shared a deep interest in natural history. Morris, born in Cork in 1810, already had some experience as an author. He had written some religious books, and had contributed articles to the journal *Naturalist*. From this time onwards, they joined forces and worked to a common purpose. Their association and friendship ended only with their deaths, within a month of each other, forty-nine years later.

The subject of their first book combined the two main interests of Morris. It was called *A Bible Natural History* and included descriptions, with illustrations engraved on wood by Fawcett, of the plants, insects, animals and birds mentioned in the Bible. The parts each contained eight of the natural history subjects and were sold for the incredibly low price of 6d each, or 10/6d for the complete volume. They followed this with a book of similar design and format, *A Book of Natural History*. Both titles had 160 hand-coloured wood-engravings, and were published by Groombridge of London. This firm had published previous books by Morris and some other authors' books on natural history. The success of these publications established a harmonious working relationship between the author, printer and publisher.

When these two early projects proved popular, they began work on their first large undertaking, *A History of British birds*. To complete the enormous task of preparing the text, and the 357 hand-coloured wood-engravings for the six volumes, they had to issue one part of four plates and twenty-four pages of letterpress every month for the next seven and a half years. This meant everyone concerned had to make a consistent effort to meet a relentless timetable. After the first four parts had been issued in 1850 the demand for the new book was so great that Fawcett could not tolerate his cramped quarters in Middle Street any longer, and so he moved to a big house called East Lodge. It had a large garden and many outbuildings in which the engraving, printing, colouring and

RED-BACKED SHRIKE

binding rooms could be accommodated, along with the fifty to sixty members of his staff.

Fawcett's aim, as stated in the Preface to his *British birds*, was to issue the work at such a price as to 'place it within the reach of every class whose taste it might happily lead them to the study of natural history.' The parts were sold at 1/- each. All of the 357 figures were engraved by Fawcett himself, then printed on greyish-lemon paper and coloured by hand. The first print of each

REDBREAST

species was carefully coloured by his wife and formed the reference copy for the colour girls. When the parts had all been issued, they were bound into six volumes on the premises at East Lodge. Bound copies sold for £5. The circulation of the book kept increasing as the work was in progress and by the time they had completed it, they had sold fifty times more than they had originally estimated. Its popular appeal was due to the modest price, the fact that it was illustrated in colour, and the discursive nature of the text.

The drawings for the bird figures were derived from a number of sources. Some of the rarer species were copied by Fawcett from Gould's books on birds, with his permission. An old schoolfellow of Morris, the clergyman-naturalist the Reverend Richard P. Alington, was the main artist and drew most of the figures. Among the attractive Alington drawings were the kingfisher, heron, moorhen, and two of the birds of prey, the hobby and kestrel. Some of the plates were unsigned, so it is not possible to know for certain who had done the original drawings.

Some of the specimens from which the drawings were made were borrowed from noblemen. Other birds were procured by clergymen, including the Rev. A. Matthew of Kilvie in Somerset and of Weston-on-the-Green, Oxford, and the Rev. John Christopher Atkinson of Kelvedon, Essex. Atkinson was a keen ornithologist and in the 1860s published his own *British birds' eggs and nests* with plates by the artist W. S. Coleman and *Sketches of natural history* with 82 figures engraved by the Dalziel Bros. after Coleman's drawings. Morris also had recourse to the numerous books on ornithology which Fawcett had collected, and mentions in the text that he had shot his own specimens on occasion, e.g. the merlin, whilst he was in Berkshire.

The bird figures were much larger than those of Bewick, occupying the whole of a quarto plate. They lost much in charm as a result. On the other hand, they were clear, large and very attractive pictures of the birds which was useful for identification purposes. The shapes were mostly correct, though the colouring was not always faithful. One unfortunate example of faulty colouring was the oystercatcher which was given a yellow beak and bright yellow legs. The foregrounds were simple, and there were no backgrounds to the illustrations. All the tree trunks, branches, and stones were done in two tones, one black and one grey, with characteristic strokes of the graver which are distinctively Fawcett's hallmark. He did not use cross-hatching, but his engraving for tone was minutely fine-lined.

The text was not always accurate, but was written in a popular style. Insufficient care had been taken in checking quite basic facts and records of the presence of rare birds were accepted uncritically. Morris was so prolific a writer that he had not sufficient time to devote to each of his books in order to make them scientifically reliable. During the course of publication of *British birds*, he was appointed Rector of Nunburnholme, another small village near Driffield. This appointment gave him fewer duties and a larger stipend, so that he had more time for his literary pursuits.

Fawcett and Morris received a number of letters from their readers asking for a companion volume to *History of British birds* dealing with the birds' eggs and nests. On the first of January 1852 they issued the first of the monthly parts of *Nests and Eggs of British Birds*. Each part sold for 1/- and had four plates. Fawcett did the engraving, and some of the drawings. He printed them on the same tinted paper that he had used for *British Birds* and then carefully touched up the illustrations by hand. The parts were bound up into three volumes with 223 plates and in his comments to the reader Fawcett included the information that 'a new invention has been applied to the department of art under which they are illustrated'. The egg-shading was done by hand, but the spots, being uniform, were most probably printed from a block. Where two flat tints were used in one egg figure, they were also printed. The new invention, therefore, was Fawcett's use of wood-engraved blocks to print in colour.

The drawings of the eggs were done, in the main, by William Richard Fisher who lived at Great Yarmouth

LITTLE
RED-BILLED
HAWK

until 1853 when he moved to London. Fisher was a personal friend of William Yarrell and had the use of Yarrell's cabinets containing the eggs of British birds. The 320 drawings by Fisher are now in the McGill University Library, Montreal. Fisher also used specimens in other naturalist's cabinets and Fawcett used specimens of some of the rarer birds' eggs sent to him by correspondents and worked from these to complete the series of drawings. The last of the parts was issued in 1856 and so *A Natural History of the Nests and Eggs* was finished before *British Birds*.

Whilst their bird books were in progress, Fawcett and the Reverend F. O. Morris were also producing a book on butterflies and issuing copies of the magazine *Naturalist*, edited by B. R. Morris from 1850–4 and F. O. Morris from 1854–6. Dr Beverley Robinson Morris was a younger brother of the Rev. F. O. Morris, and qualified as a doctor of medicine. He also took a great interest in the natural history of this country until he departed for America in 1856. The previous year Fawcett had engraved the illustrations for sixty coloured wood-engravings illustrating B. R. Morris' *British Game Birds and Wildfowl*. These plates were hand-coloured, formed a double-page spread and were separately hinged. Usually, one bird was depicted, occasionally a pair. The portraits were beautifully executed and skilfully coloured. Some were signed 'B. Fawcett del', for example the woodcock, barbary partridge and little bustard. The text gave details of distribution, habits, value as food and a description of the bird, its nest and eggs. The monthly parts, costing 2/6d, were eventually bound into one volume.

In 1853 Fawcett became acquainted with another very popular author of bird books, a chemist by profession. Henry Gardiner Adams also recognised that there was a current demand for attractively produced books on butterflies, shells and birds. Some of his titles were for children e.g. *Humming-birds*, 1856 and *Nests and Eggs of Familiar Birds* published in 1854, each with eight plates printed by Fawcett. He also printed eight plates in 1853 to illustrate Adams' *Favourite Song Birds*. This had been issued in 1851 with 12 lithographs after drawings by Edward Gilks and published by W. S. Orr. Groombridge published the later editions in which the eight chromoxylographs depicted a blackbird, nightingale, skylark, bullfinch, chaffinch, song thrush, wren and redstart. The printed wood-engravings are delicately coloured with smooth flat tints and are most attractive.

SKYLARK

Fawcett also did the drawings and engraved the blocks for the 238 plates depicting the *Birds of Europe not observed in the British Isles*, printed 1859–63. The text was written by Dr Charles Bree and complemented *History of British Birds* by Morris. It was also the only book on the subject, at a reasonable price, for many years. About sixty 1/- parts were published by Groombridge.

The individual technique which Fawcett had developed for printing in colour was a remarkable breakthrough in colour-printing. He engraved on Turkish boxwood for the key and colour blocks using extremely fine-lined engraving. Very occasionally, when a surface of perfectly flat colour was required, he used either stereo metal or a Persian boxwood—some of these exceptions may be seen in the plates depicting eggs in the Morris book about British birds' eggs. Only hand-presses were used in the printing, eight being in operation when work was at full pressure. The chromoxylographs were passed to the colour room for final touching by hand and then to the binding room. Fawcett always inspected every plate when printed, in order to maintain his high standards.

Fawcett's outstanding pupil, Alexander Francis Lydon was fully familiar with all these processes. Frank, as he was called, was so skilled at his work, both drawing and engraving that Fawcett put him into each department to learn the job from start to finish. Beilby had done this with Bewick, Baxter with Harrison Weir, and it is a measure of the ability of each of these pupils that they were treated with such respect and consideration by their masters.

Frank Lydon was particularly good at water-colour drawing. This was to be his main contribution to Fawcett's workshop for many years after Lydon was out of his apprenticeship. Though Fawcett was a keen painter, and very good at drawing, when the pressure of business increased, he gradually left this side of the business to Lydon. The young artist had done all the drawings for the 79 coloured and 909 monochrome illustrations in E. J. Lowe's *A Natural history of ferns, British and exotic* (published 1856–60). The coloured illustrations would have required a number of blocks for each design, and this would entail the preparation of well over a thousand blocks for this title alone, though we know that it was not the only book on which Lydon was employed at that time. Fawcett had, once again, found a workman of great skill combined with phenomenal output.

Lydon helped Fawcett with four bird books, during the years 1870–1888. They all contained beautiful chromoxylographs. In 1871 Hugh Piper issued two tiny books, one on *Pigeons* and one on *Poultry*. There was a growing interest in keeping and breeding pigeons, pheasants and poultry. Much of this interest had been generated by W. B. Tegetmeier and was stimulated by the enthusiasm for pigeon breeding and racing of King Edward VII when Prince of Wales. Lydon did a frontispiece for each of Piper's books, depicting many species on each plate.

Fawcett and H. G. Adams were joined by H. B. Adams in 1874 to produce an extremely attractive little book about the *Smaller British birds*. It was originally issued in parts, then bound into a single volume containing 16 plates of birds and 16 plates of eggs. The plates are signed 'A. F. Lydon' and form lovely small pictures with the birds well drawn and the colouring good. Of course, the subjects themselves are among our most appealing species—tits, wagtails, pipits, larks, chats, finches, flycatchers, but Lydon's treatment adds considerably to their charm. Each plate had a number of species from the various groups or families. These are some of the best chromoxylographs of birds and it is a truly delightful book. It was not published by Groombridge, but by G. Bell & Sons.

Fawcett suffered a most severe blow in the 1870s when Groombridge & Sons found themselves financially embarrassed. The transactions between them had run into several thousand pounds. They had made a great success of their enterprises, but competition was building up from other publishers of ever-cheaper illustrated books produced by other methods of colour printing. Fawcett's financial loss was heavy. He turned to G. Bell & Sons for publishing his subsequent books, though still retaining the engraving and actual printing of the text and illustrations in his own workshop.

The last bird book for which Lydon did the drawings and Fawcett the printing was Dr William Thomas Greene's *Parrots in captivity*, published in 1884–8 in three volumes and two parts of a proposed fourth volume. A. F. Lydon was the artist for 90 chromoxylographs, whilst Dr Greene's text covered some 80 species of parrots, parakeets, macaws, lories and cockatoos. A black outline plate was lightly printed, and the delicate colours printed on top. The plates depicted one bird, or a pair, with a branch and foliage; an example is printed as the colour frontispiece to this book. Dr Greene dedicated the three volumes to Fawcett.

For some reason, never stated or explained, Lydon decided to leave Fawcett in 1883—in the middle of preparing Dr Greene's *Parrots*. Lydon had been with Fawcett for over thirty years. All accounts of the two men pass quietly over this episode which must have been painful for both of them and could not have happened at a worse time for Fawcett. Not only had he found increasing competition from other printers, and lost the sympathetic and excellent workmen at Groombridge's establishment, but he had been suffering from loss of sight due to a cataract on one eye. He had this cataract removed when he was over seventy years of age.

The volume of Fawcett's business, of necessity, lessened. Both he and Morris were now elderly, but in 1892 they brought out a new edition of *British Birds* which had been out of print for some time, adding 29 new species. A new edition of *Nests and Eggs* came out about the same time. The following January, 1893, Fawcett died, and Morris died in February. The business was carried on for a little while by one of Fawcett's sons but it did not flourish and was sold in January 1895 by auction. The whole of the printing plant was bought by Messrs W. Phelps & Son of Birmingham.

Lydon, on leaving Driffield, went to London and established himself there as an illustrator. He worked for another thirty years. Besides his work on the design of plates for books, he drew and exhibited many pictures. He also did hundreds of illustrations for periodicals including *Animal World, Canary and Cage Bird Life, The Fancier's Gazette, Illustrated London News, Poultry* and many others. For the *Feathered World* alone he contributed nearly 1,000 drawings of birds which were reproduced in monochrome and colour. He maintained a high standard of bird portraiture during the course of his very long career. Of the later books which he illustrated, the most notable is Dr R. B. Sharpe's *Sketch-book of British birds* with 305 colour printed illustrations with the text run round. Twenty-five of Lydon's excellent wood-engravings were used for Dr Greene's *Birds of the British Empire* 1898. This was the largest number by one artist in this book to which at least a dozen other people contributed. Lydon also did some of the drawings for W. Swaysland's *Familiar wild birds*, including some egg figures, but these were reproduced by chromolithography.

Lydon, Morris, Groombridge, Ridley, and many more, had liked and respected Fawcett and worked with him for a long period. All of them must have worked very long hours. Most probably, he had their loyalty because Fawcett's interest in his work-people extended to their personal lives as well as their careful training and supervision during work time. Between them they produced many titles with beautiful illustrations, maintaining a high standard of production despite their tremendous output. What is even more commendable, is that these standards were consciously kept as an objective whilst they were aiming at producing books which the poorer members of society could afford. Cheapness combined with excellence was their aim, and their remarkable achievement.

*Bird books illustrated by Fawcett*

ADAMS, H. G. Favourite song birds; containing a popular description of the feathered songsters of Britain. 1853 8 wood-engravings
Later editions with chromoxylographs.
Artist: Edward Gilks
Engraver and printer: B. Fawcett

ADAMS, H. G. The Smaller British birds with descriptions of their nests, eggs, habits . . . 1874 front. and 32 chromoxylogs. Reissue 1894
Artist: A. F. Lydon
Engraver and printer: B. Fawcett

BREE, C. R. A History of the birds of Europe, not observed in the British Isles 1859–63 4 vols with 238 chromoxylogs.
2nd ed 1875–6 5 vols with 253 plates
Artist and engraver and printer: B. Fawcett

GREENE, W. T. Birds I have kept in years gone by, with original anecdotes and full directions for keeping them successfully. 1885. 16 plates chromoxylogs.
Artists: B. Fawcett; A. F. Lydon
Engraver and printer: B. Fawcett and A. F. Lydon

GREENE, W. T. Parrots in captivity 1884–5 3 vols + parts I and II of vol. 4. 90 chromoxylogs.
Artist: A. F. Lydon
Engraver and printer: B. Fawcett

MORRIS, B. R. A History of British game birds and wild fowl 1855 60 partly col. printed and partly hand-col. wood-engravings
2nd ed 1873; 3rd ed 1881
4th ed revised by W. B. Tegetmeier 1895 2 vols.
Artist: B. Fawcett (some signed by him; some unsigned)
Engraver and printer: B. Fawcett

MORRIS, F. O. A History of British birds 1851–7 6 vols with 358 hand-col. wood-engravings. New issue 1856–62
2nd ed 1870–1 (the best edition) 365 col. pl. in 6 vols
3rd ed 1891–4 (29 new species added)
4th ed 1895–6 with 394 plates
5th ed revised and brought up to date by H. K. Swann 1903 400 plates
Artists: Rev. R. P. Alington; B. Fawcett (after Gould)
Engraver and printer: B. Fawcett
Colourer: Martha Porter (later Mrs Fawcett)

MORRIS, F. O. A Natural history of the nests and eggs of British birds 1852–6 3 vols with 232 coloured wood-engravings
Many new issues and editions
Artists: W. Fisher; B. Fawcett
Engraver and printer: B. Fawcett

MORRIS, F. O. A Book of natural history, containing a description of animals and birds, trees, plants, insects . . . mentioned in the Holy Scriptures, illustrated by coloured engravings 1849 166 hand-col. wood-engravings
Artist and engraver and printer: B. Fawcett

PIPER, H. Pigeons 1871 1 chromoxylog.
Artist: A. F. Lydon
Engraver and printer: B. Fawcett
PIPER, H. Poultry 1871 1 chromoxylog.
Artist: A. F. Lydon
Engraver and printer: B. Fawcett

# *A Change of Style*

The style and design of modern wood engraved birds is different from that of Bewick and his successors. This interesting change of style is to some extent due to the introduction of the camera. Field photographers can now supply authors with scientifically accurate pictures of birds to be used for identification purposes and this rôle is no longer required of wood engraved figures. The wood engraver, free from the stylistic constraints imposed by designing a long series of illustrations of the bird and branch type, can now depict birds in flight, preening, sleeping, bathing, even scratching. It is no longer a matter of concern if the picture of the active bird obscures an identification mark on the wing, or a foot or claw, since the more relaxed wood engraved illustrations are used to decorate and enliven the text and not to demonstrate field or species recognition marks.

Creative wood-engraving, as opposed to the use of wood-engraving merely for diagrams, was little used in bird books for three decades, but was revived shortly before the first world war and flourished after 1920 when the Society of Wood Engravers was founded. One of the leaders of the movement was Robert John Gibbings who executed the wood-engravings of birds for Godfrey P. Harrison's *A Bird Diary* published in 1936 and for *The Charm of Birds* 1927 by Viscount Grey of Fallodon, besides many bird figures in general natural history books.

The birds in these books are designed to illustrate a discursive text, and this is also true of the type of book which other bird artist-engravers of this century have illustrated. As a result, we find many more pictures of birds in flight, and groups of inter-acting birds such as three or four gulls fighting for food. Birds within their environment, rather than superimposed on small backgrounds, is another feature of modern illustration.

There is also a new artistic element, for the illustration is designed to complement the text and type. The freer style is better suited to the descriptive text, and the engraving is prepared and drawn after the artist has read through the book (or written it himself) and been prompted and inspired to create an illustration pertinent to an idea in the text. Chronicles of bird life and selective discourses on special groups of birds, for example, birds of the seashore, are ideal subjects for this type of illustration.

One of several bird artist-engravers was Eric Fitch Daglish. Ater six years in the British Army, followed by two years as scientific correspondent to the Evening Standard, he found his true vocation when he became an author and engraver in 1925. He joined the Society of Wood Engravers in 1924, the same year in which he became interested in wood engraving and cut his first block. Thereafter he did many wood-engravings of natural history subjects. He illustrated his own book

'*Woodcuts of British Birds with descriptions . . .*' with 20 plates in 1925, and *How to see Birds*, again with 20 engraved plates in 1932. Daglish also did some engravings for D. Dewar's *Game Birds* 1928, and E. M. Nicholson's edition of Gilbert White's *Selborne* published by Thomas Butterworth in 1929. In this particular *Selborne* the wood-engravings were printed on separate sheets of paper and mounted onto pages of the book. A thick black line round the square area of each illustration detracted from the charm of the plates, nine of which depicted birds. E. M. Nicholson's *Birds in England* had eight wood-engravings by Daglish when first published in 1926.

In 1948 Daglish produced *Birds of the British Isles described and engraved with 48 wood engravings by the author.* Twenty-five of these engravings were hand-coloured for a special limited edition of 1,500 copies. It is one of the very few bird books published in the 20th century with hand-coloured plates.

Today, it is very rare, however, for an engraver's wooden block actually to be used for printing. Although the hand-process of wood-engraving still survives, the mechanical process of photography nowadays intervenes between the craftsman engraver and the printer.

The outstanding wood-engraver of birds in the present century is Charles F. Tunnicliffe who has illustrated many of his own delightful books. These include *Shorelands Summer Diary* 1951 with 114 wood-engravings, *Birds of the Estuary* 1952 and *Mereside Chronicle* 1947. An early work by Mary Priestley called *A Book of Birds* which was published by Gollancz in 1937 also contained some of Tunnicliffe's careful, accurate engravings of birds, three of which are reproduced here.

Tunnicliffe has created some truly beautiful pictures and studies of birds. These are in the best tradition,

WHITE STORKS

combining accuracy with artistic feeling, and vivacity with veracity. He was the worthy recipient of the R.S.P.B. gold medal for services to bird protection in 1975, and the subject of an appreciative article in the *Birds* magazine November-December of that year. He was awarded the O.B.E. in 1978.

His first painting for the R.S.P.B. was used on a Christmas card in 1950. He has been a regular contibutor to the Royal Academy exhibitions and was elected an Associate in 1944 and Royal Academician ten years later.

SNIPE

His acute observation and deep knowledge of the structure of birds is revealed in his book *Bird Portraiture* (published in 1945). A charming account of his own early life on a farm and training as an artist and engraver at the Royal College of Art in South Kensington is given in *My Country Book* (another 'Studio' publication, issued in 1942). Both of these books, beautifully illustrated, prove Tunnicliffe to have a deep respect for, and love of, nature. These attitudes are reflected in the honest and very successful attempts to reproduce its subjects in his paintings and wood engravings, and mark him as the outstanding bird artist engraver of the first three-quarters of the 20th century.

FALCON

# *Bibliography*

## General Reference Books

ALEXANDER, W. B. Ornithological illustration *in* Endeavour 1953, 12:144–153

ANKER, J. Bird books and bird art. 1938.

BALL, A. & MARTIN, M. The Price guide to Baxter prints, 1974.

BLAND, D. A History of book illustration. 1974.

BRYAN, M. A Biographical dictionary of painters and engravers: edited by G. C. Williams, 1903–05 5 vols (1st ed 1816 2 vols).

BURCH, R. M. Colour printing and colour printers. 1911.

DALZIEL, G. & E. The Brothers Dalziel: a record of 50 years' work 1840–90. 1901.

DICTIONARY OF NATIONAL BIOGRAPHY: edited by Sir L. Stephens and Sir S. Lee. 1882.

FISHER, S. W. English watercolours. 1970.

GRAVES, A. The British Institution 1806–1867: a

complete dictionary of contributors and their work from the foundation of the institution. 1875.

GRAVES, A. A Dictionary of artists who have exhibited works in the principal London exhibitions 1760–1893. 1895.

GRAVES, A. Royal Academy of Arts: a complete dictionary of contributors and their work 1769-1904. 1905–06.

GRAVES, A. Society of Artists of Great Britain 1760–1791: Free Society of Artists 1761–1783: a complete dictionary of contributors and their work 1760–1791. 1907.

JACKSON, J. & CHATTO, W. A. A Treatise on wood engraving. 1839.

JOHNSON, J. Works exhibited at the Royal Society of British Artists 1824–1893. 1975 2 vols.

KNIGHT, D. M. Natural science books in English 1600–1900. 1972.

LEWIS, C. T. C. The Story of picture printing in England during the 19th century. 1927.

LINDLEY, K. The Woodblock engravers. 1971.

MacLEAN, R. Victorian book design and colour printing. 1963.

MITCHELL, Sir P. C. Centenary history of the Zoological Society of London. 1929.

MULLENS, W. H. & SWANN, H. K. A Bibliography of British ornithology from the earliest times to the end of 1912, including bibliographical accounts of the principal writers and bibliographies of their published works. 1917

NICHOLSON, E. M. Birds in England. 1926.

NISSEN, C. Die illustrierten Vogelbücher: ihre Geschichte und Bibliographie. 1953.

RONSIL, R. Bibliographie ornithologique française. 1948–9 2 vols.

SAWYER, F. C. A Short history of the libraries and list of MSS and original drawings in the British Museum (Natural History). Bulletin of the B.M. (Nat. Hist.) Historical Volume 4 no. 2. 1971.

SITWELL, S. and others Fine bird books 1700–1900 by Sacheverell Sitwell, James Fisher and H. Buchanan. 1953.

SKETCHLEY, R. E. D. English book illustrators of today. 1903.

SWAINSON, W. Taxidermy, with the biographies of zoologists and notices of their works. 1840.

WOOD, C. A. An Introduction to the literature of vertebrate zoology based chiefly on the titles in the Blacker Library of Zoology, the Emma Shearer Wood Library of Ornithology, the Bibliotheca Osleriana and other libraries of the McGill University, Montreal. 1931.

ZIMMER, J. T. Catalogue of the Edward E. Ayer ornithological library, Field Museum, Chicago. 1926 2 vols.

SIX-WIRED BIRD OF PARADISE

EAGLE OWL

# *Biographical Sources*

## *HISTORY AND DEVELOPMENT*

BALL, A. & MARTIN, M. The Price guide to Baxter prints. 1974.

DALZIEL, G. & E. The Brothers Dalziel: a record of 50 years' work 1840–90. 1901.

DEWAR, D. William MacGillivray *in* Scottish Field 1958 May: 65–71.

LODGE, G. *In* Ibis 1954:474.

MILLAIS, J. G. *In* Who Was Who 1929–40; *in* Ibis 1931:589; *in* British Birds 1931, 24:362–3.

MULLENS, W. H. William MacGillivray and William Yarrell *in* British Birds 1909, 2:389–399.

PALMER, A. H. The Life of Joseph Wolf. 1895.

SWAINSON, W. Taxidermy, with the biographies of zoologists and notices of their works. 1840.

THOMSON, A. A Biographical account of the late William MacGillivray *in* Edinburgh New Philosophical Journal 1853, 54:189–206.

THOMSON, J. A. The Life of William MacGillivray. 1910.

*Thomas BEWICK*

BELL, J. G. A Descriptive and critical catalogue of works illustrated by Thomas and John Bewick, wood engravers of Newcastle-upon-Tyne; with an appendix of their miscellaneous engravings, brief sketches of their lives, and notices of the pupils of Thomas Bewick. 1851.

BEWICK T. Letters to Dovaston; edited by G. Williams. 1968.

BEWICK, T. Memoirs of Thomas Bewick; edited by Ian Bain. 1975.

COPELAND's ornaments. H. Copeland and Mathias Lock were responsible for most of the designs in Chippendale's famous trade catalogue, *The Gentleman's & Cabinet-maker's Director*, pubd in 1754.

JACKSON, J. & CHATTO, W. A. A Treatise on wood engraving. 1839.

JARDINE, Sir W. Naturalist's Library, vol. 15. Memoir of Bewick, written by Wm Turner.

MULLENS, W. H. Thomas Bewick and George Montagu *in* British Birds 1908–09, vol. 11:351–361.

RAYNER, S. Wood engravings by Thomas Bewick. 1946.

REYNOLDS, G. Thomas Bewick: a résumé of his life and work. 1949.

RUSH, J. The ingenious Beilbys. 1973.

THOMSON, D. C. The Water colour drawings of Thomas Bewick. 1930.

*Thomas C. EYTON*

BAGSHAW, Sam. History, gazetteer and directory of Shropshire. 1851.

EDDOWES Salopian Journal 27 Oct 1880:7 Death of T. C. Eyton of Eyton.

EYTON, T. C. Letters of T. C. Eyton to Wm Yarrell 1833–40 *in* Balfour Library, Zool. Dept. Univ. of Cambridge.

EYTON, T. C. A Catalogue of the species of birds in his possession. 1857.

EYTON, T. C. Osteologia Avium, 1871–8 Supp. 1869 Supp. 11, 1873–5.

IBIS 1908, Supp: 179 T. C. Eyton.

*William YARRELL*

GAGE, A. T. History of the Linnean Society of London. 1938.

JENYNS, L. Reminiscences of William Yarrell. 1885.

LINNEAN SOCIETY Proceedings 1858, II:xxxiii–xxxviii Obit. of Yarrell.

MULLENS, W. H. William MacGillivray and William Yarrell *in* British Birds 1909, 2:389–399.

YARRELL, W. History of British Fishes 3rd ed 1859, includes an account of the life of William Yarrell by Sir John Richardson.

YARRELL, W. Letters to Sir William Jardine *in* Zoology Library, B. M. (Nat. Hist.).

*William DICKES*

BOOKSELLER Christmas number, 10 December 1864.

DOCKER, A. The Colour prints of William Dickes. 1924.

LEWIS, C. T. C. The Story of picture printing in England during the 19th century. 1927.
MacLEAN, R. Victorian book design and colour printing. 1963.

*Benjamin FAWCETT*

BOOKSELLER 1893 March 7:17 The late Mr Benjamin Fawcett of Driffield.
LEWIS, C. T. C. The Story of picture printing in England during the 19th century. 1927.
MORRIS, M. C. F. Benjamin Fawcett, colour printer and engraver. 1925.
MORRIS, M. C. F. Memoir of F. O. Morris. 1897.
RICHARDSON, E. W. A Veteran naturalist: the life and work of W. B. Tegetmeier. 1916.

*A CHANGE OF STYLE*

EMPSON, P. Wood engravings by Robert Gibbings. 1959.
HADFIELD, J. A Master of wood engraving (Robt Gibbings) *in* Country Life, Feb. 4, 1960:217.
ROYAL ACADEMY OF ARTS, *Diploma Galleries.* (Catalogue of) Bird drawings by C. F. Tunnicliffe R.A. 3 Aug.–29 Sept 1974.
R.S.P.B. Portrait of the master (C. F. Tunnicliffe) *in* Birds Nov-Dec. 1975:24–28.
TUNNICLIFFE, C. F. My country book. 1942. (autobiographical notes inter-alia).
WHO WAS WHO 1961–70: 271 Eric Fitch Daglish.
WHO'S WHO 1978 Charles F. Tunnicliffe.

# *Sources of the Illustrations*

Coloured frontispiece: W. T. Greene, Parrots in Captivity.

pp. 3–7, 9, 133: A. H. Evans, Vol IX Cambridge Natural History series, Birds.

p. 10: Gilbert White, The Natural History and Antiquities of Selborne, with notes by Edward Turner Bennett and others.

pp. 19–20: J. G. Wood, The Illustrated Natural History, Birds.

p. 21: C. A. Johns, British Birds in their Haunts.

pp. 24–25: W. Swaysland, Familiar Wild Birds.

p. 26: William Yarrell, A History of British Birds, 4th edition.

p. 29: Newcastle upon Tyne City Library.

pp. 30–57: Thomas Bewick, A History of British Birds, editions 1–6.

p. 58: Thomas Bewick, A General History of Quadrupeds.

pp. 61–70: T. C. Eyton, A History of the Rarer British Birds.

pp. 73–90: William Yarrell, A History of British Birds, editions 1–4.

pp. 93, 94, 95, 97, 98, 104: Blaine, Rural Sports.

pp. 96, 100, 101: Anon, British Birds, Water Birds.

pp. 99: A. Pratt, Our Native Songsters.

pp. 107, 108, 111, 112: F. O. Morris, A History of British Birds.

p. 115: C. R. Bree, A History of the Birds of Europe, not observed in the British Isles.

p. 117: H. G. Adams, Favourite Song Birds.

p. 124: Godfrey Harrison, A Bird Diary.

pp. 128–130: Mary Priestley, A Book of Birds.

p. 134: Gilbert White, The Natural History and Antiquities of Selborne, with notes by Edward Turner Bennett and others.

# Index